THE
DELL BOOK
OF
LOGIC
PROBLEMS #4

THE
DELL BOOK
OF
LOGIC
PROBLEMS #4

Editor-in-Chief • Erica L. Rothstein
Special Editor • Kathleen Reineke
Editor • Theresa Turner

A DELL TRADE PAPERBACK

A DELL TRADE PAPERBACK

Published by
Dell Publishing
a division of
Bantam Doubleday Dell Publishing Group, Inc.
666 Fifth Avenue
New York, New York 10103

ISBN: 0-440-50181-4

Printed in the United States of America
Published simultaneously in Canada

December 1989

10   9   8   7   6   5   4   3   2   1

MV

A WORD ABOUT THIS BOOK

Lt. Columbo, Jessica Fletcher, move over!

Yes, you heard it right, both of you. You're not the only ones who can put all the clues together and come up with the correct answer. A whole new crop of sleuths is in the making, and they won't play second fiddle to anyone.

Even though all you Logic Problem solvers might never make your way into starring roles on your own television series, you can still join the ranks of the great detectives, past and present. And while the Logic Problem kind of detecting won't lead to your own Private Investigator's license, either, we can promise it will indeed lead to many diverting hours of puzzle solving. That's true whether you're an old hand at Logic Problems from years of solving in the Dell Puzzle Magazines or previous editions of THE DELL BOOK OF LOGIC PROBLEMS, or picking up this book for a very first go-round.

Contained in this book are 75 brand-new Logic Problems—and you don't have to be a master detective to master all of them. (For some of the Challengers, it probably wouldn't be too bad if you *were* a master!) As a matter of fact, the easy puzzles which begin on page 25, are designed specifically for those solvers who are new to Logic Problems. These puzzles will help a new solver to acquire and hone the basic skills of solving, while providing the more experienced with a quick solve to whet their appetite for the more challenging ones that follow.

Also included in this book, beginning on page 13, is a comprehensive, step-by-step, how-to guide which is not just for beginners—even veteran solvers can benefit from some of the hints contained therein. In this section, solvers at all levels of skill will find invaluable information on the use of solving charts and additional insight into the thought processes that go into solving Logic Problems. We urge you to take a look.

THE EDITORS
Dell Book of Logic Problems #4
Dell Publishing
245 Park Avenue
New York, New York 10167

A LETTER FROM THE EDITOR

It is my pleasure to welcome you to the fourth edition of THE DELL BOOK OF LOGIC PROBLEMS. Like its predecessors, this book contains 75 never-before-published puzzles which have been chosen especially for your solving enjoyment. The Logic Problems contained herein have all been created by the world's best constructors in order to assure you the ultimate puzzle entertainment.

Do you ever wonder how constructors keep coming up with new situations for their puzzles? It never ceases to amaze me! We've asked, of course; most say inspiration comes from the newspaper or TV or from everyday experiences. But clearly some puzzles result from out-of-the-ordinary inspirations. One puzzle in particular comes to mind here—"Alien Luv"—which you will find on page 94. Perhaps when the constructor returns from her "intergalactic tour," she can enlighten us on the experiences from which she draws her ideas. But, seriously, I hope you enjoy the amusing twist to the answer of that particular puzzle.

Much has been done to assure you hours of challenging solving in the following pages. Each puzzle has been carefully selected, edited, trial solved, reworked, rewritten, retyped, retrial solved, etc., etc., all to make certain that you get the best solve possible. Many hands worked hours, days, and weeks to bring you these puzzles. Two (perhaps that should be four?) in particular, Stacey E. Feinman and Victor Mather, solved away diligently in this group effort to make these puzzles everything you could expect from Dell. Editors couldn't do all this work if they didn't love it, because it simply can't be done well if not done with concern for the solver uppermost. Think it's hard to figure out how to solve a Logic Problem? Just imagine, then, the difficulty editors have when they know a puzzle *doesn't* work, and they have to figure out not only what's wrong, but how to fix it. And then, start all over again to make sure it works!

Well, care we do, about every single aspect of the puzzles collected here. We hope it shows (it's supposed to!) But even more importantly, we hope you enjoy this latest edition of THE DELL BOOK OF LOGIC PROBLEMS.

KATHLEEN REINEKE
Special Editor
Dell Puzzle Publications

CONTENTS

EASY LOGIC PROBLEMS

MEDIUM LOGIC PROBLEMS

HARD LOGIC PROBLEMS

CHALLENGER LOGIC PROBLEMS

HOW TO SOLVE LOGIC PROBLEMS

For those of you who are new to Logic Problems: On the following pages you will find some insights into the thought processes that go into solving these puzzles, as well as detailed instructions on the use of charts as solving aids. We suggest you scan these instructions to familiarize yourself with the techniques presented here. Whenever you feel that you're ready to try your hand at solving, turn to the first puzzle (which you will find on page 25) and dig right in. If, even after you have studied these instructions, you should find yourself stuck while solving, turn to the solution for that puzzle and try to follow the reasoning given there. The solutions are not just a listing of "who did what," but rather a step-by-step elimination of possibilities, which you should find invaluable on your journey along the road to mastery of Logic Problems.

The 75 Logic Problems in this book are just that—problems based on logic, to which you need bring no specialized knowledge or extensive vocabulary. Instead, all you will need is your common sense, some reasoning power, and a basic grasp of how to use the charts or other solving aids provided. The problems themselves are all classic deduction problems, in which you are usually asked to figure out how two or more sets of facts relate to each other—what first name belongs with which last name, for example. All of the facts you will need to solve each puzzle are always given.

The puzzles are mostly arranged in increasing order of difficulty—the first few are rather easy to solve, then the puzzles get more difficult as you continue through the book. The final puzzles are especially challenging. If you are new to Logic Problems, we suggest that you start with the first puzzles, progressing through the book as you get more expert at solving.

Of the three examples which follow, the first is, of course, the most basic, but the skills utilized there will help you tackle even the most challenging challenger. Example #2 will help you hone those skills and gives valuable hints about the use of a more complicated chart as a solving aid. The third member of the group will introduce those puzzles for which the normal solving chart is not applicable. You will notice that in each of these examples, as in all the Logic Problems in this book, the last part of the introduction will tell you what facts you are to establish in solving that puzzle. Now, if you are ready to begin, read through the introduction and the clues given with Example #1.

EXAMPLE #1

A young woman attending a party was introduced to four men in rather rapid succession and, as usual at such gatherings, their respective types of work were mentioned rather early in the conversation. Unfortunately, she was afflicted with a somewhat faulty memory. Half an hour later, she could remember only that she had met a Mr. Brown, a Mr. White, a Mr. Black, and a Mr. Green. She recalled that among them were a photographer, a grocer, a banker, and a singer, but she could not recall which was which. Her hostess, a fun-loving friend, refused to refresh her memory, but offered four clues. Happily, the young woman's logic was better than her memory, and she quickly paired each man with his profession. Can you? Here are the clues:

1. Mr. White approached the banker for a loan.

2. Mr. Brown had met the photographer when he hired him to take pictures of his wedding.

3. The singer and Mr. White are friends, but have never had business dealings.

4. Neither Mr. Black nor the singer had ever met Mr. Green before that evening.

	Black	Brown	Green	White
banker				
grocer				
photo.				
singer				

You know from the last part of the introduction what it is you are to determine—you are to match each man's last name with his profession. The chart has been set up to help you keep track of the information as you discover it. We suggest that you use an X in a box to indicate a definite impossibility and a • (dot) in a box to show an established fact.

Your first step is to enter X's into the chart for all of the obvious possibilities that you can see from information given in the clues. It is apparent from clue 1 that Mr. White is not the banker, so an X would be entered into the White/banker box. Clue 2 makes it clear that Mr. Brown is not the photographer, so another X in the Brown/photographer box can be entered. Clue 3 tells you that Mr. White is not the singer. And from clue 4 you can see that neither Mr. Black nor Mr. Green is the singer. Each of these impossibilities should also be indicated by X's in the chart. Once you have done so, your chart will look like this:

	Black	Brown	Green	White
banker				X
grocer				
photo.		X		
singer	X		X	X

Remembering that each X indicates that something is *not* a fact, note the row of boxes at the bottom—corresponding to which of the men is the singer. There are four possibilities, and you have X's for three of them. Therefore, Mr. Brown, the only one left, has to be the singer. Put a dot (•) in the singer/Brown box. Also, remember that if Mr. Brown is the singer, he is not the photographer (which we knew, we have an X); and he cannot be the grocer or the banker either. Thus, you would put X's in those boxes too. Your chart would now look like this:

	Black	Brown	Green	White
banker		X		X
grocer		X		
photo.		X		
singer	X	•	X	X

Now you seem to have a "hopeless" situation! You have used all the clues, and you have matched one man with his profession—but the additional X's entered in the chart do not enable you to make another match, since the possibilities have not been narrowed down sufficiently. What to do next?

Your next step is to reread the clues, at the same time considering the new information you have acquired: You know that Mr. Brown is the singer and that he has done business with the photographer (clue 2). But the singer has never done business with Mr. White (clue 3) or with Mr. Green (clue 4). And that means that neither Mr. White nor Mr. Green can possibly be the photographer. You can now place X's in those boxes in the chart. After you have done so, here is what you will have:

	Black	Brown	Green	White
banker		X		X
grocer		X		
photo.		X	X	X
singer	X	•	X	X

And you see that you do have more answers! The photographer must be Mr. Black, since there are X's in the boxes for the other names. Mr. White, also, must be the grocer, since there is an X in the other three boxes under his name. Once you have placed a dot to indicate that Mr. Black is the photographer and a dot to show that Mr. White is the grocer (always remembering to place X's in the other boxes in the row and column that contain the dot) your chart will look like this:

	Black	Brown	Green	White
banker	X	X		X
grocer	X	X	X	•
photo.	•	X	X	X
singer	X	•	X	X

You can see that you are left with one empty box, and this box corresponds to the remaining piece of information you have not yet determined—what Mr. Green's profession is and who the banker is. Obviously, the only possibility is that Mr. Green is the banker. And the Logic Problem is solved!

Most of the Logic Problems in this book will ask you to determine how more than two sets of facts are related to each other. You'll see, however, that the way of solving a more involved Logic Problem is just the same as Example #1—*if* you have a grasp of how to make the best use of the solving chart. The next example of a Logic Problem is presented in order to explain how to use a bigger chart. As before, read through the problem quickly, noting that the introduction tells you what facts you are to determine.

EXAMPLE #2

Andy, Chris, Noel, Randy, and Steve—one of whose last name is Morse—were recently hired as refreshment vendors at Memorial Stadium; each boy sells only one kind of fare. From the clues below, try to determine each boy's full name and the type of refreshment he sells.

1. Randy, whose last name is not Wiley, does not sell popcorn.

2. The Davis boy does not sell soda or candy.

3. The five boys are Noel, Randy, the Smith boy, the Coble boy, and the boy who sells ice cream.

4. Andy's last name is not Wiley or Coble. Neither Andy nor Coble is the boy who sells candy.

5. Neither the peanut vendor nor the ice cream vendor is named Steve or Davis.

	Coble	Davis	Morse	Smith	Wiley	candy	ice.	pean.	pop.	soda
Andy										
Chris										
Noel										
Randy										
Steve										
candy										
ice.										
pean.										
pop.										
soda										

Note that the chart given is composed of three sets of boxes—one set corresponding to the first and last names; a second set (to the right) corresponding to first names and refreshment; and a third set, below the first set, corresponding to the refreshment and last names. Notice, too, that these sets are separated from each other by heavier lines so that it is easier to find the particular box you are looking for.

As in Example #1, your first step is to enter into the boxes of the chart the impossibilities. Keep in mind that you have many more boxes to be concerned with here. Remember, ROW indicates the boxes that go horizontally (the Andy row, for example) and the word COLUMN indicates the boxes that go vertically (the Coble column, for instance).

Clue 1 tells you that Randy's last name is not Wiley, and Randy does not sell popcorn. Thus, enter an X into the Randy/Wiley box and another X in the Randy/popcorn box in the Randy row. Clue 2 says that the Davis boy sells neither soda nor candy. Find Davis and go down that column to the Davis/soda box and put an X in it; then find the Davis/candy box in that same column and place an X in that box.

Clue 3 tells you a few things: It gives you all five of the boys, either by his first name (two of them), his last name (another two of them), or by what refreshment he sells (the remaining boy). You then know something about all five—one boy's first name is Noel, another's is Randy; a third boy has the last name Smith, a fourth has the last name Coble; and the fifth sells ice cream. All of these are different people. So, in the chart you have a lot of X's that can be entered. Noel's last name is neither Smith nor Coble, so enter X's in the Noel/Smith, Noel/Coble boxes; nor can Noel be the ice cream seller, so put an X in the Noel/ice cream box. Randy is neither Smith nor Coble, and Randy does not sell ice cream, so put the X's in the Randy/Smith, Randy/Coble, and Randy/ice cream boxes. And neither Smith nor Coble sells ice cream, so enter an X in those two boxes.

16

Clue 4 tells you that Andy's last name is neither Wiley nor Coble. It also says that Andy does not sell candy and neither does the Coble boy. By now you probably know where to put the X's—in the Andy/Wiley box, the Andy/Coble box, the Andy/candy box, and in the box in the Coble column corresponding to candy. From clue 5 you learn that neither Steve nor Davis is the boy who sells either peanuts or ice cream. (One important point here—read clue 5 again, and note that this clue does *not* tell you whether or not Steve's last name is Davis; it tells you only that neither the peanut seller nor the ice cream vendor has the first name Steve or the last name Davis.) Your chart should now look like this:

	Coble	Davis	Morse	Smith	Wiley	candy	ice.	pean.	pop.	soda
Andy	X				X	X				
Chris										
Noel	X			X			X			
Randy	X			X	X		X		X	
Steve							X	X		
candy	X	X								
ice.	X	X		X						
pean.		X								
pop.										
soda		X								

From this point on, we suggest that you fill in the above chart yourself as you read how the facts are established. If you look at the Davis column, you will see that you have X's in four of the refreshment boxes; the Davis boy is the one who sells popcorn. Put a dot in the Davis/popcorn box. Now, since it is Davis who sells popcorn, none of the other boys does, so you will put X's in all of the other boxes in that popcorn row.

Your next step will be to look up at the other set of refreshment boxes and see what first names already have an X in the popcorn column. Note that Randy has an X in the popcorn column (from clue 1). Thus, if you know that Randy does not sell popcorn, you now know that his last name is not Davis, since Davis is the popcorn seller. You can then put an X in the Randy/Davis box. After you've done this, you'll see that you now have four X's for Randy's last name. Randy has to be Morse, the only name left, so enter a dot in the Randy/Morse box. Don't forget, too, to enter X's in the boxes of the Morse column that correspond to the first names of the other boys.

Now that you know Randy is Morse, you are ready to look at what you've already discovered about Randy and transfer that information to the Morse column—remember that since Randy is Morse, anything that you know about Randy is also true of Morse, as they're the same person. You'll see that an X for Randy was entered from clue 3: Randy does not sell ice cream. Then Morse cannot be the ice cream seller either, so put an X in the Morse column to show that Morse doesn't sell ice cream.

Once the Morse/ice cream X is in place, note what you have established about the Wiley boy: His is the only last name left who can sell ice cream. Put the dot in the Wiley/ice cream box and enter X's in the Wiley column for all the other refreshments. Your next step? As before, you are ready to determine what this new dot will tell you, so you will go up to the other set of refreshment boxes and see what you have established about the ice cream vendor. He's not Noel or Steve—they have two X's already entered in the chart. Now that you have established the Wiley boy as the ice cream seller, you know that his first name can't be either Noel or Steve because neither of those boys sells ice cream. Once you've put X's in the Noel/Wiley box and the Steve/Wiley box, you'll see that you know who Wiley is. Remember that clue 4 had already told you that Andy's last name is not Wiley, so you have an X in the Andy/Wiley box. With the new X's, do you see that Wiley's first name has to be Chris? And since Chris is Wiley, and Wiley sells ice cream, so, of course, does Chris. Thus, you can put a dot in the Chris/ice cream box. And don't forget to put X's in the Chris row for the other refreshments and also in the ice cream column for the other first names.

Notice that once Chris Wiley is entered in the chart, there are now four X's in the Coble column, and Steve is the one who has to be the Coble boy. Put in the dot and then X's in the Steve row, and your chart looks like this:

	Coble	Davis	Morse	Smith	Wiley	candy	ice.	pean.	pop.	soda
Andy	X		X		X	X	X			
Chris	X	X	X	X	•	X	•	X	X	X
Noel	X		X	X	X		X			
Randy	X	X	•	X	X		X		X	
Steve	•	X	X	X	X		X	X		
candy	X	X			X					
ice.	X	X	X	X	•					
pean.		X			X					
pop.	X	•	X	X	X					
soda		X			X					

See that there are four X's in the Smith/first name column, so Smith's first name must be Andy. And Noel's last name is Davis, because he's the only one left. Remember— look down the Davis row and see—we already know Davis sells popcorn. So, Noel, whose last name is Davis, sells popcorn. And, of course, there should be X's in all the other boxes of the Noel row and the popcorn column.

Now that you have completely established two sets of facts—which first name goes with which last name—you can use the two sets of refreshment boxes almost as one. That is, since you know each boy's first name and last name, anything you have determined about a first name will hold true for that boy's last name; and, naturally, the reverse is true: whatever you know about a boy's last name must also be true of that boy's first name.

For example, you know that Coble is Steve, so look down the Coble column and note that you have already put X's in the candy, ice cream, and popcorn boxes. Go up to the Steve row and enter any X's that you know about Coble. After putting an X in the Steve/candy box, you'll see that you've determined that Steve sells soda. As always, don't forget to enter X's where appropriate once you've entered a dot to indicate a determined fact. These X's are what will narrow down the remaining possibilities.

Things are really moving fast now! Once you've entered the appropriate X's in the Steve row and the soda column, you will quickly see that there are four X's in the candy column—so, Randy (Morse) is the candy vendor. By elimination, Andy (Smith) sells peanuts and this Logic Problem is completely solved.

Many of the Logic Problems in this book will have charts that are set up much like the one in Example #2. They may be bigger, and the puzzle may involve matching more sets of facts, but the method of solving the Logic Problem using the chart will be exactly the same. Just remember:

Always read the whole problem through quickly. What you are to determine is usually stated in the last part of the introduction.

When using solving charts, use an X to indicate a definite impossibility and a • (dot) to indicate an established fact.

Once you have placed a dot to indicate an established fact, remember to put X's in the rest of the boxes in the row and the column that contains the dot.

Every time you establish a fact, it is a good idea to go back and reread the clues, keeping in mind the newly found information. Often, you will find that rereading the clues will help you if you seem to be "stuck." You may discover that you *do* know more facts than you thought you did.

Don't forget, when you establish a fact in one part of a solving chart, check to see if the new information is applicable to any other section of the solving chart—see if some X's or dots can be transferred from one section to another.

Just one other note before we get to Example #3, and this note applies to both the most inexperienced novice and the most experienced expert. If ever you find yourself stymied while solving a problem, don't get discouraged and give up—turn to the solution. Read the step-by-step elimination until you get to a fact that you have not established and see if you can follow the reasoning given. By going back and forth between the clue numbers cited in the solution and the clues themselves, you should be able to "get over the hump" and still have the satisfaction of completing the rest of the puzzle by yourself. Sometimes reading the solution of one puzzle will give you important clues (if you'll pardon the pun) to the thought processes involved with many other puzzles. And now to the last of our trio of examples.

Sometimes a Logic Problem has been created in such a way that the type of chart you learned about in Example #2 is not helpful in solving the problem. The puzzle itself is fine, but another kind of chart—a fill-in type—will better help you match up the facts and arrive at the correct solution. Example #3 is a puzzle using this type of solving chart.

EXAMPLE #3

It was her first visit home in ten years, and Louise wondered how she would manage to see her old friends and still take in the things she wanted to in the seven days she had to spend there. Her worry was needless, however, for when she got off the plane Sunday morning, there were her friends—Anna, Cora, Gert, Jane, Liz, and Mary— waiting to greet her with her seven-day visit all planned. The women knew that Louise wanted to revisit the restaurant where they always used to have lunch together, so Louise's vacation began that Sunday afternoon with a party. After that, each of the women had an entire day to spend with Louise, accompanying her to one of the following things: a ball game, concert, the theater, museum, zoo, and one day reserved for just shopping. From the clues below, find out who took Louise where and on what day.

1. Anna and the museum visitor and the woman whose day followed the zoo visitor were blondes; Gert and the concertgoer and the woman who spent Monday with Louise were brunettes. *(Note: All six women are mentioned in this clue.)*

2. Cora's day with Louise was not the visit that occurred the day immediately following Mary's day.

3. The six women visited with Louise in the following order: Jane was with Louise the day after the zoo visitor and four days before the museumgoer; Gert was with Louise the day after the theatergoer and the day before Mary.

4. Anna and the woman who took Louise shopping have the same color hair.

	Monday	Tuesday	Wednesday	Thursday	Friday	Saturday
friend						
activity						

As before (and always) read the entire puzzle through quickly. Note that here you are to determine which day, from Monday to Saturday, each woman spent with Louise and also what they did that day. The solving chart, often called a fill-in chart, is the best kind to use for this puzzle. You won't be entering X's and dots here; instead, you will be writing the facts into the chart as you determine them and also find out where they belong.

From clue 1 you can eliminate both Anna and Gert as the woman who took Louise to the museum and the concert. And neither of these activities took place on a Monday, nor did Anna or Gert spend Monday with Louise. You have discovered some things, but none of them can yet be entered into the chart. Most solvers find it useful to note these facts elsewhere, perhaps in the margin or on a piece of scratch paper, in their own particular kind of shorthand. Then when enough facts have been determined to begin writing them into the chart, you will already have them listed.

Do you see that clue 2 tells you Mary did not see Louise on Saturday? It's because the clue states that Cora's day was not the visit that occurred immediately following Mary's day, and thus, there had to be at least one visit after Mary's. You still don't have a definite fact to write into the chart. Don't lose heart, though, because . . .

. . . clue 3 will start to crack the puzzle! Note that this clue gives you the order of the six visits. Since the days were Monday through Saturday, the only possible way for Jane to be with Louise the day after the zoo visitor and four days before the museumgoer is if the zoo visit took place on Monday, Jane was with Louise on Tuesday, and the museumgoer was with Louise on Saturday. These facts can now be written into the chart—Monday zoo, Tuesday Jane, Saturday museum. Three days have been accounted for. The last part of clue 3 gives you the other three days: with Wednesday, Thursday, and Friday still open, the theatergoer must be the Wednesday friend, Gert is the day after, or Thursday, and Mary saw Louise on Friday. These facts, too, should be written in the chart. Once you've done so, your chart will resemble this one:

	Monday	Tuesday	Wednesday	Thursday	Friday	Saturday
friend		Jane		Gert	Mary	
activity	zoo		theater			museum

Now go back to clue 1 and see what other facts you can establish. There are three blondes—Anna, the museum visitor, and the woman whose day followed the zoo visitor's. The chart shows you that this last woman was Jane. From clue 4 you learn that the woman who took Louise shopping and Anna have the same color hair—blond. The woman who took Louise shopping is not Anna (they're two separate people), nor is she the museum visitor, so she must be the woman whose day followed the zoo visitor's, Jane. That fact can be written in the chart.

You can also, at this point, establish what day Anna spent with Louise. Since you know it's not Monday (clue 1) and Anna is not the museumgoer (also clue 1), the only day left for her is Wednesday, so Anna took Louise to the theater. Clue 2 tells you that Cora's day did not immediately follow Mary's, so Cora's day can't be Saturday, and must be Monday. By elimination, Liz (listed in the introduction) spent Saturday with Louise at the museum.

It may be helpful to make a note of the hair colors mentioned in clue 1, perhaps under the relevant columns in the chart. These hair colors can again be used at this point. We've now established the blondes as Anna, Jane, and Liz; the brunettes are Gert, the concertgoer, and Cora. The only possibility is that Mary is the concertgoer. Everything has now been determined except what Gert did, so, by elimination, Gert must have taken Louise to a ball game (from the introduction).

	Monday	Tuesday	Wednesday	Thursday	Friday	Saturday
friend	Cora	Jane	Anna	Gert	Mary	Liz
activity	zoo	shopping	theater	ball game	concert	museum
	bru	blo	blo	bru	bru	blo

Are all Logic Problems easy to solve? No, of course not. Many of the puzzles in this book are much more complicated than the three examples and should take a great deal more time and thought before you arrive at the solution. However, the techniques you use to solve the puzzles are essentially the same. All the information needed to solve will be given in the puzzle itself, either in the introduction or the clues. As you eliminate possibilities, you will narrow down the choices until, finally, you can establish a certainty. That certainty will usually help narrow down the possibilities in another set of facts. Once you have determined something, you will probably need to return to the clues and reread them, keeping in mind what facts you have now established. Suddenly a sentence in the clues may tell you something you could not have determined before, thus narrowing down the choices still further. Eventually you will have determined everything, and the Logic Problem will be solved.

EASY LOGIC PROBLEMS

1 FOUR A'S ANTIQUES

Ms. Kent and three associates own and operate the Four A's Antiques Shop. Each woman is an expert in a different type of antiques. From the following clues, determine each woman's full name and area of expertise, along with the order in which the four arrived at their shop yesterday.

1. Amy arrived before Ms. Trent (who is not the furniture expert), who arrived before the jewelry expert.

2. Anna arrived after the clothing expert but before Ms. Dent.

3. Ada arrived before Ms. Brent.

4. Ava's specialty is antique books.

The solution is on page 133.

The solution is on page 133.

	Brent	Dent	Kent	Trent	books	clothing	furniture	jewelry	1	2	3	4
Ada												
Amy												
Anna												
Ava												
1												
2												
3												
4												
books					
clothing					
furniture					
jewelry					

2 IT'S DONE WITH MIRRORS

by Mary Marks Cezus

Four friends went to the Mirror Palace yesterday and each purchased a mirror which had a different shape and a different price. With some reflection you should be able to use the clues below to determine each friend's first name, last name, mirror shape, and mirror price ($175, $225, $300, or $350).

1. Oscar's mirror cost less than Evans' mirror, but more than the oval one.

2. The round mirror cost less than Nicole's mirror but more than Downing's mirror.

3. Myra (who is not Glass) and the person who bought the hexagonal mirror are both saving their money for another mirror purchase soon.

4. Lloyd and Downing both bought some special glass cleaner with their mirror purchases.

5. "It's mere money," responded the purchaser of the hexagonal mirror when Evans expressed concern about the prices.

6. The person who bought the rectangular mirror (who is not Myra) paid less than Fairfax.

7. Both Downing and the person who bought the oval mirror will put their mirrors in their bedrooms.

The solution is on page 133.

You might find it helpful to use both the regular chart and the fill-in chart provided below in solving this problem.

	Downing	Evans	Fairfax	Glass	hex.	oval	rect.	round	175	225	300	350
Lloyd												
Myra												
Nicole												
Oscar												
175												
225												
300												
350												
hex.												
oval												
rect.												
round												

first name	last name	shape	price

3 ALL THE KING'S HORSES

by W. H. Organ

All Mr. King's horses were gentle creatures. This made them very popular with the girls at the nearby Montclair School for Young Ladies, who took riding lessons several times each week. On one recent afternoon, Kay and five of her schoolmates took their usual lesson, each one on her favorite horse (one was named Angel). From the following clues can you name each girl, her horse, its distinctive color or marking, and its age? (None of the horses was over 6 years or under 3.)

1. Prince, a sorrel gelding, fell while attempting a jump. Sue, who was riding Duke, also took a spill; neither the horses nor the riders were hurt.

2. Eve rode an Appaloosa.

3. Thor, a black stallion, was twice as old as Nan's pinto filly.

4. Pat's Pixie and Tomboy are both 4-year-olds.

5. Two of the horses are 5-year-olds.

6. Bea did not ride the black horse.

7. The chestnut is older than the white horse.

The solution is on page 133.

The solution is on page 133.

rider	horse	color	age

4 MEDITERRANEAN WINDS

by W. H. Organ

Dick and four other young seamen were gathered around a table at a Boston quayside inn discussing the weather—not ordinary weather, but the sudden and often violent winds that are encountered in the Mediterranean. The men all served on different ships and each of their ships had suffered wind damage on their last cruise to the Mediterranean. From the following clues can you determine the full names of the seamen, the type of ship in which each served, the name of the wind which had caused the damage (one was called a lebeche), and the port they went to for repairs?

1. Adam's ship and the ship damaged by the mistral were brigantines; Price's ship and the one which went to Tripoli for repairs were both brigs.

2. One of the ships, a bark, lost a foresail in a sudden gale called a bora. It went to Split for repairs.

3. Smith, whose first name is not Adam, was on the ship that went to Marseilles for repairs.

4. Tom Davis' ship weathered a sirocco with only minor damage.

5. Newton's ship is not a brigantine. His first name is not Hiram.

6. Neither Hiram nor Abel served in a brig.

7. Jones's ship was repaired at Barcelona.

8. The ship damaged by the wind called a gregale was repaired at Valetta.

The solution is on page 133.

first name	last name	type of ship	type of wind	port

5 COUNTY FAIR COWS

by Varda Maslowski

The cows shown at this year's county fair included many fine animals. Rainbow and three others won the top four prizes. From the following clues, match each prize-winner with its owner (one owner's first name is Cal; one's last name is McKay), and the prize it won.

1. Sal's cow (which is not Lightning) placed higher than Thunderbolt, but not as high as Mr. McGee's cow.

2. Al's cow placed one place higher than Hal's, but not as high as Mr. Magee's.

3. Blue Streak, which is not owned by Mr. McGee, placed either first or last.

4. Mr. McFee is not Hal.

The solution is on page 134.

prize	first name	last name	animal's name

6 HAWAIIAN HOLIDAY

by Mary Marks Cezus

Brett, Jennifer, and two other friends are enjoying a vacation in the Hawaiian islands. They, however, did not plan their vacations together and were surprised when they ran into each other at Honolulu International Airport a few days ago. They each booked with a different tour group and tomorrow all will be participating in different activities on different islands. Using the clues below can you determine each vacationer's name, activity (one is shopping), island (one is Oahu), and tour group (one is Carvas)?

1. The last time the sunbather traveled to Hawaii he took Worldwide Tours and loved it. He was sorry to find all their tours full when he tried to make a reservation this time.

2. The woman who is going snorkeling and the man who will be on Kauai agreed to meet Lauren for lunch when they return to the mainland.

3. The person with Multitour and the person who will be on the Big Island traveled to Bermuda together last year.

4. The man who will be on Maui borrowed a guidebook from the person on Tampos Tours.

5. The traveler who came to sightsee chose the Tampos Tour on a previous trip to Hawaii and did not have enough time to see everything, so did not choose that group again.

6. The person with Multitour and Michael have never been to Hawaii before this trip.

7. The person going to the Big Island is not with the Worldwide group.

The solution is on page 134.

vacationer	activity	island	tour group

7 BIRTHDAY GREETINGS

by Mary A. Powell

For Steve Webber's 50th birthday, his wife and children arranged a surprise party for him. They invited relatives, old and new friends, and even former teachers. Five of the people invited (including Steve's first grade teacher) were unable to attend, but sent their greetings by telephone. From the following clues, can you find the full names (one last name is Denver) of the absent well-wishers, their relation to Steve, and the state from which each called?

1. The call from Florida came before the call from Mr. Wilcox, who is not Harry.

2. Steve's niece called from California; an uncle called from Alaska; and his 90-year-old grandmother, who had a bad cold, called from her home in the next town.

3. Leona, who is not Grayson, called after Michael.

4. The man who was Steve's first roommate in college called from Hawaii; that call came after the call from McMillan, who is not Jane.

5. The three calls from relatives were the one from Alice, the one from Peterson, and the one from Ohio.

The solution is on page 134.

first name	last name	relation	state

8 MUSIC LESSONS

by Varda Maslowski

Every Saturday, Mrs. Green gives music lessons to Ted and three other adult students. Each studies a different instrument; one studies the violin. From the following clues, determine each student's full name (one's last name is Freeman) and the instrument he or she studies, along with the order in which the four receive their lessons.

1. John receives his lesson before Frohman and after the guitar student.

2. Elaine receives her lesson before Mr. Freedman and after the piano student.

3. Neither Friedman nor Nancy is Mrs. Green's first student.

4. John is not the viola student.

The solution is on page 135.

The solution is on page 135.

	first name	last name	instrument
1st			
2nd			
3rd			
4th			

9 THE KNITTERS

by Anne Smith

Betty and four other women of Craftsville recently decided to learn how to knit. For her first project, each woman made a different item and each used a different color of yarn. From the following clues, determine the full name of each woman (one surname is Graves), the item each woman knitted (one woman knitted a scarf), and the color of each item (one was yellow).

1. The mittens were green.

2. Three of the women, Anna Franklin, Ms. Jones, and the one who learned how to make a sweater, took knitting lessons at a local craft store; the other two, Ms. Iverson and the one who made the red item (who is not Edith), taught themselves how to knit.

3. Clara and the woman who made the white item had been friends since grade school.

4. Donna made the blue blanket.

5. Clara took knitting lessons, but she did not learn how to make a sweater.

6. Ms. Jones and the woman who knitted the hat didn't know each other before they met at their knitting class.

7. Ms. Harris made her item from red yarn.

The solution is on page 135.

knitters		color	item

VALENTINE WEDDINGS

by Lois Bohnsack

Four couples sent their engagement notices to the local newspaper. The women's editor remarked to her assistant, "What a coincidence! Each of these weddings is planned for Valentine's Day. The last names of the grooms are McDowell, McGill, McKinney, and McNeal. The last names of the brides are Dowell, Gill, Kinney, and Neal. However, no engaged couple has similar names." From the clues below can you determine each person's full name and correctly identify the engaged couples?

1. Tammy, Susan, and Ms. Gill work together.

2. Ms. Neal is not engaged to Keith.

3. Neither Patty nor Ruth is Mr. McKinney's fiancee.

4. Mr. McDowell (who is not marrying Patty) is Jerry's neighbor.

5. Mr. McGill, Larry, and Mike are salesmen.

6. Tammy and Ms. Neal are planning church weddings. Neither is engaged to Jerry or Mr. McKinney.

7. Ruth and Larry are not engaged to each other.

The solution is on page 135.

	McD	McG	McK	McN	Patty	Ruth	Susan	Tammy	Jerry	Keith	Larry	Mike
									engaged to			
Dowell												
Gill												
Kinney												
Neal												
Jerry												
Keith												
Larry												
Mike												
Patty												
Ruth												
Susan												
Tammy												

engaged to

11 GREETING CARD PARTNERS

by Varda Maslowski

Four friends joined forces to create a greeting-card company. Each performs a different function in this partnership (one is the shipper), and each has a different number of years of experience. From the following clues, determine the full name (one's first name is Harold), nickname, and function of each partner, along with the number of years of experience each one has (one partner has twenty years' experience).

1. Eugene has half as much experience as "Chip" (who is not Mr. Mercer) but twice as much as Mr. Malden.

2. Mr. Mason has as much experience as two other partners—Barry and "Flip"—put together.

3. Arthur has four years' experience.

4. "Buzz" has more experience than Mr. Moore and at least one other partner.

5. Neither "Flip" nor "Tank" is the designer.

6. Mr. Mercer has more experience than Arthur but not as much as the printer, who is not Mr. Moore.

7. Eugene does not have as much experience as Barry.

8. Mr. Malden is not the packager.

The solution is on page 136.

	Malden	Mason	Mercer	Moore	Buzz	Chip	Flip	Tank	des.	pack	print	ship	experience least			most
Arthur																
Barry																
Eugene																
Harold																
experience least																
experience most																
des.																
pack																
print																
ship																
Buzz																
Chip																
Flip																
Tank																

12 MILE RELAY TEAM

by Lois Bohnsack

Gary, Henry, Ivan, and Jeff, the members of the Central High School mile relay team, broke the previous record at the State Track Meet this past spring with a total time of 3 minutes and 22 seconds. One boy ran his leg of the race in 48 seconds; the other times were 49, 52, and 53 seconds. From the clues below can you determine each boy's name (the last names are Robbins, Shelton, Tyler, and Utley), his time, and his position (i.e. first leg, second leg, etc.)?

1. The four boys are: the boy who ran in 52 seconds, the boy who ran the third leg, Gary, and Utley.

2. The first runner of the relay team was not the fastest, but he was faster than Henry.

3. Tyler ran before Robbins; neither is Jeff.

4. The second runner was not as fast as Gary.

5. Utley and the runner of the first leg are seniors. Ivan, whose time was slower than the third runner, is a junior.

6. The boy who ran in 52 seconds and the boy who ran the second leg are also on the cross-country team.

The solution is on page 136.

The solution is on page 136.

	runner		time
1st leg	_____	_____	_____
2nd leg	_____	_____	_____
3rd leg	_____	_____	_____
4th leg	_____	_____	_____

13 THE PINEVALE BUGLE

by W. H. Organ

During the past week the Pinevale Bugle published six letters to the Editor. The writers were all concerned about a local problem: how forty acres of land bordering the north side of the city should be zoned. Each letter expressed a different opinion. The letters were published on successive days, starting on Monday and continuing through Saturday. From the following clues, can you determine the full names of the letter writers (one surname is Curtin), the use to which each one felt the land should be put, and the day of the week each letter was published?

1. Sarah and the person who favored the land being used for industrial purposes have been residents of Pinevale for over fifteen years. King and the one whose letter was published on Saturday have been residents for less than ten years.

2. Rose Dorgan's letter, which favored residential zoning, was the third to be published.

3. Jackson wanted the land to be used for a city park. His letter was published later in the week than Brown's, but a day earlier than King's.

4. Dolores favored agricultural zoning.

5. Fred's letter was published on Thursday.

6. Floyd, whose letter was published before Lopez's, thought the land should be reserved for much-needed schools.

7. Dan's surname is not Brown.

8. The letter in the Monday paper favored commercial zoning.

The solution is on page 136.

	letter writer		land use
Mon.	_____	_____	_____
Tues.	_____	_____	_____
Wed.	_____	_____	_____
Thur.	_____	_____	_____
Fri.	_____	_____	_____
Sat.	_____	_____	_____

MEDIUM LOGIC PROBLEMS

14 EASTER EGG HUNT

by Elaine M. Decker

At the annual Smallville Easter egg hunt, the top five egg hunters were all different ages. From the following clues, can you determine each child's full name (one first name is Laura; one last name is Farmer), his or her age, and the number of eggs each child found?

1. None of the children have the same first and last initials.

2. The number of eggs found by the Smith child was twice his or her age.

3. Chris, who is not the Grant child, is twice as old as the Lawson child, who is one year younger than Greg.

4. Stephen, who is not Lawson, found three eggs more than the six-year-old; Stephen found three eggs less than the Grant child.

5. The Star child, who is the oldest, found one egg less than the eight-year-old.

6. The seven-year-old found fifteen eggs.

7. Mandy found thirteen eggs.

The solution is on page 137.

The solution is on page 137.

	Farmer	Grant	Lawson	Smith	Star	# of eggs					ages				
Chris															
Greg															
Laura															
Mandy															
Stephen															
ages										
# of eggs					

15 RETRAINING

by Tara Lynn Fulton

Sylvia and four other workers in Midsville were unemployed for a short time last year when they decided to change their occupations (one was a telephone operator) and undergo retraining for new jobs. The five are now happily re-employed (one is a mechanic). From the clues below, can you determine each worker's full name (one surname is Swanson), former job, and present job?

1. The five workers are Tom, the former welder, the present arcade manager, Ms. Cortez, and the present fitness instructor (who is not Mr. Bertram).

2. Ralph used to be a foreman.

3. Marie, who is neither Cortez nor Monroe, used to be a secretary.

4. Mr. Hampton is now a mail carrier.

5. The programmer, who is not Erica, used to repair TVs.

The solution is on page 137.

former job	name	present job
_____	_____	_____
_____	_____	_____
_____	_____	_____
_____	_____	_____
_____	_____	_____

16 TRIMMING THE TREE

by Julie Spence

While trimming their Christmas tree last December, the four Smith children, including Steve and Susan, discussed their favorite things about Christmas, including tree trimmings. From the clues below, can you determine each child's age, favorite color of Christmas lights (one likes gold), and favorite tree ornament (one likes the star which goes on top of their tree)? Note: All the ages are whole numbers of years.

1. The seven-year-old isn't the youngest.

2. The child who likes the cardinal ornament doesn't like blue or red lights.

3. The boy who likes green lights is two years younger than Sally, who is twice as old as the girl whose favorite ornament is a Santa's elf.

4. The one who likes a snowman ornament is younger than the one who likes blue lights.

5. Sam is three years older than the boy who likes red lights.

The solution is on page 138.

	cardinal	elf	snowman	star	blue	gold	green	red	ages		
Sam											
Sally											
Steve											
Susan											

ages								

	cardinal	elf	snowman	star
blue				
gold				
green				
red				

WOODLAND WATERING SPOT

by Diane C. Baldwin

Four different groups of woodland creatures—including pumas, partridges, raccoons, and ravens—came early one morning to drink from the spring in the clearing. All the groups were different sizes, numbering from two to five creatures, and each group came a different way and at a different time, so as not to be seen by the others. From the information below, can you tell the order in which the animals came to drink, how many were in each group, and which way each approached the spring?

1. The largest group didn't come by way of the old oak tree.

2. The ones who appeared out of the tall grass numbered the same as their order of arrival.

3. The total number of birds was three times the number of animals in one of the other groups.

4. The first group to arrive was the smallest; it was neither the group which came by way of the path nor the one which came by way of the old oak tree.

5. The raccoons, who weren't the creatures who came over the rock, weren't last.

6. Those whose names began with the same letter didn't arrive consecutively.

The solution is on page 138.

18 MOTHER'S DAY GIFTS

by Diane Yoko

For Mother's Day last year, Tom and his four friends all emptied their piggy banks and went to a nearby variety store to buy their mothers gifts; one child bought nail polish. From the following clues, can you determine each child's full name and age and the gift each chose for his or her special mom?

1. Carl is one year younger than Dotty, who is one year younger than the Brook child, who isn't Rusty.

2. The Smith child is older than the one who bought the bubble bath.

3. The Heart child bought a puzzle magazine.

4. Rusty is one year older than the girl who bought the candy bar but one year younger than the Polar child, who is ten.

5. The Jones child is two years younger than Beth.

6. Beth bought neither the chocolate-covered cherries nor the candy bar.

The solution is on page 138.

The solution is on page 138.

child	age	gift

46

19 THE COFFEE SHOPPERS

by W. R. Pinkston

When a new store opened featuring worldwide coffees, Joe was the first customer. He was followed by five other customers, and each bought a variety from a different country. From the following clues, you are to determine each shopper's full name (one surname is Long), the coffee variety each chose (one is Lintong), and the coffee's country of origin (one is Kenya).

1. Meg isn't Ms. King.

2. One variety cost less than Leo's purchase, and only one cost more than the strong variety bought by Beth.

3. Leo was the only man to select a strong variety such as Kalossi or another from Sumatra; he purchased neither of these.

4. Neither Sue nor Smith bought the most expensive kind, which is from Celebes.

5. The Peaberry variety was bought by a man; the one from Celebes was bought by a woman.

6. Mr. Frey's choice was from Java; Leo's was not from Java or China.

7. Ms. Ward is the only woman who selected a mild variety; she didn't get the mild Colombian blend or the milder Estate variety.

8. The man who bought the least expensive Excel variety is neither Ian nor Mr. Dunn.

9. One of the least expensive coffees is the Yunnan variety, which isn't from Sumatra.

The solution is on page 139.

shopper		coffee	country

47

20 AIRPORT ACQUAINTANCES

by Varda Maslowski

When Mr. Owens and four other travelers found themselves fogged in at the airport, they decided to wait at the airport's coffee shop for the fog to lift. While talking at the table (see diagram), they learned that each of them was going to a different destination; one was headed to college in Pennsylvania. From the following clues, determine each traveler's full name (one's first name is Gertrude), destination, and seat number at the table.

1. The five included: Karen, Mr. Quinlan, Ira, the traveler going on vacation to Hawaii, and the business traveler going to New York.

2. Mills sat directly across from another traveler.

3. Jim sat directly across from Nolan.

4. A woman sat in Seat #3.

5. Both the business traveler going to Alaska and the traveler going on vacation to Colorado each sat next to only one person.

6. Price was on a business trip.

7. Mills' destination was east of the Mississippi.

8. Karen was on a vacation trip.

9. Jim sat immediately to the left of Hank.

The solution is on page 139.

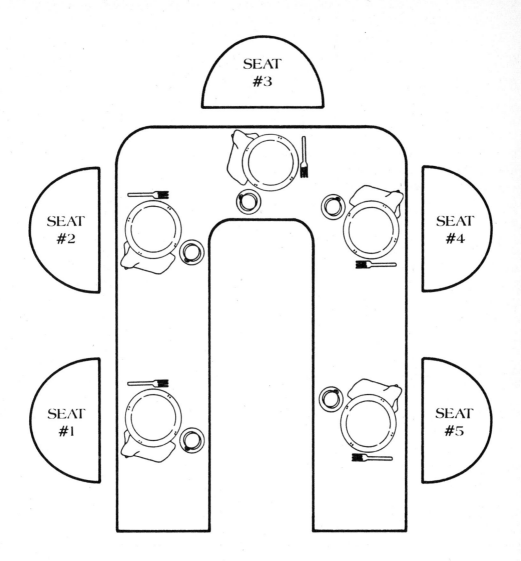

21 SHOWTIME

by Anne Smith

Six married couples attended a show last Saturday evening. Each couple went to see a different show at a different theater. From the following, determine the full name of each couple (one wife was Vanessa, one husband was Frank, and one surname was Taylor), the theater each couple attended, and the time each show began. (Each woman has the same surname as her husband.)

1. Rita, Teresa, the Hugheses, and Doug attended different shows at the local mall, which were showing (in no particular order) at Cinemas 1, 2, 3, and 4; the Kings and Ursula went to comedy shows at the Uptown Theater and the Bijou, respectively.

2. Bob, Chuck (who saw a horror show), the Greens, and Sally (who went to the mall) each went to a different theater for the 7 p.m. show; Andrew and the Joneses saw different shows at 9 p.m.

3. Of the four couples going to the mall, the number of the cinema Wanda attended was lower than either Andrew's or Rita's, but higher than the number of the cinema attended by the Sampsons.

4. Ed attended a show at Cinema 4.

5. There were two 7 p.m. and two 9 p.m. shows at the mall.

The solution is on page 139.

The solution is on page 139.

CINEMA 1	CINEMA 2
_____	_____
time	time
_____ and _____	_____ and _____
_____	_____

CINEMA 3	CINEMA 4
_____	_____
time	time
_____ and _____	_____ and _____
_____	_____

BIJOU	UPTOWN
_____	_____
time	time
_____ and _____	_____ and _____
_____	_____

22 THE ENLISTEES

by W. H. Organ

The three Martin youngsters and the three Fremont youngsters, upon graduating from Harbor View High School over a four-year period, enlisted in various branches of the armed forces; one of the Martins was the last of the six to enlist. While in the service, they all took advantage of the opportunity to obtain vocational training, and on completion of their enlistments, they returned to Harbor View—where, as a result of their training and experience, they had no difficulty in finding employment. From the following clues, can you determine each graduate's name, branch of service, and employment after returning to Harbor View?

1. John and the young man who went into the Marine Corps were first to enlist, the year before the two graduates who joined the Army; none of these four is the one who found work in radio repair.

2. Reginald, who did not serve in the Army, found employment as an automobile mechanic.

3. Two of the graduates enlisted in the Navy; Dana was not one of them.

4. Willy, who is Pat's brother, found a job as a carpenter; Pat served in the Air Force and is not the radio repairer or the dental technician.

5. One of the six found employment as a stenographer; her brother got a job as an electrician.

6. One of the young women found employment as a dental technician.

7. Kim was not the last of the six to enlist.

The solution is on page 140.

enlistee	service	employment
_____	_____	_____
_____	_____	_____
_____	_____	_____
_____	_____	_____
_____	_____	_____
_____	_____	_____

23   MIXED-UP COFFEE MUGS

Three male employees of the Moneyfell Investors' Corporation (Maxwell, Chad, and George) and three of their female co-workers, including Phyllis, share a six-space coffee-cup holder at their workplace, as shown in the diagram. One day, the night janitor decided to be extra kind and wash each of the six mugs. Unfortunately, he did not pay very close attention to where the cups went and managed to replace them so that not even one was returned to its correct place. From the clues below, can you determine where the six cups (which included a mug adorned with a rainbow) were when picked up the next morning by the six employees, as well as where each actually belonged?

1. The cup bearing the initial of its owner should hang in the row directly above the plain-colored cup, but it ended up on the bottom row.

2. The mug which had a wise saying printed on it was picked up by a man, but it actually belonged to a woman.

3. Julie, who doesn't really wake up until after drinking her coffee, had drunk an entire cup before she realized she had Donna's mug.

4. The mug which was adorned with the logo of its owner's college alma mater was switched kitty-cornered from where it actually belonged.

5. The striped mug ended up directly below where it should be.

6. The cups belonging to two of the men were switched so that each ended up with the other's cup.

The solution is on page 140.

52

DONNA

JULIE

GEORGE

MAXWELL

PHYLLIS

CHAD

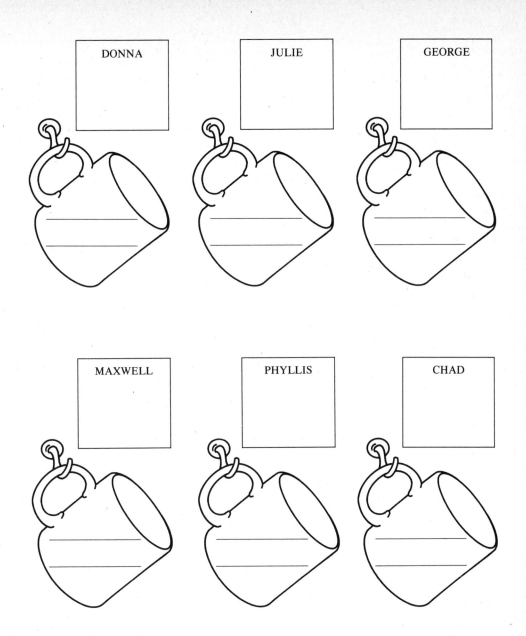

RELICS OF THE PAST

by W. R. Pinkston

The Abbott County Historical Society had not accomplished very much until a kind donor gave them an acre of land containing an ancient home and barn. This initiated restoration work by volunteers along with requests for local people to donate items of the past that would be appropriate for the old house and barn. Within two weeks, six items were received from people in various towns. From the following clues, you are to determine the order in which items were donated, the donors (one surname was Olson), and the hometowns (one was Canton) of the donors.

1. The Midway resident delivered his gift which was followed by one more before the Fenton man's gift, which wasn't the fifth.

2. Two of the three women gave household items the second week, and one was the first of three items received that week.

3. The wood-burning kitchen stove was donated just after Ms. Beebe's gift, which wasn't the antique sewing machine from Keeton.

4. The old wagon wasn't the first gift, but it was donated just before Mr. Jacob's gift, which was just ahead of the trunk.

5. Ms. Alcott's donation came just after Mr. Reyes', which was just after one from the Senton woman, who isn't Ms. Miller.

6. The water wheel arrived just before a donation from the Fulton resident, which was just before the harness set.

The solution is on page 140.

	donor	hometown	item
1			
2			
3			
4			
5			
6			

25 ADDLED AIRPORT

by Claudia Strong

Martha Longworth and five other members of her family had flights out of the Addled Airport between 5:00 and 6:00 one evening. Among the family members who traveled, there were two pairs and two who traveled singly, going on four different flights (one was Flight #211) on four different airlines (including Icarus, Inc.). None of the flights left at exactly the same time, and each flight left from a different one of Addled Airport's four concourses, which are lettered A through D. From this and the clues that follow, can you find the flight number, airline, concourse, and exact take-off time for each of the Longworth family members?

1. As they left each other in the airport's main lobby, three people took the shuttle car toward Concourses A and B; the other three took the walkway toward Concourses C and D.

2. Egbert's flight (which was not #409) left 10 minutes after Georgette and her sister departed on Friendly Bird Airways.

3. Flight #148 departed later than the flight departing from Concourse C.

4. Ada was glad that her flight number was less than 400, since that meant she did not have to take a shuttle car to her concourse.

5. The first family members to take off were the pair who did so at exactly 5:15; the Buy & Fly flight was the next to leave, exactly ten minutes later.

6. Fred, who did not fly alone, left 25 minutes earlier than the flight departing from Concourse B.

7. The pair who flew on Flight #760 departed after Mary and the Non-Skeddo pair.

The solution is on page 141.

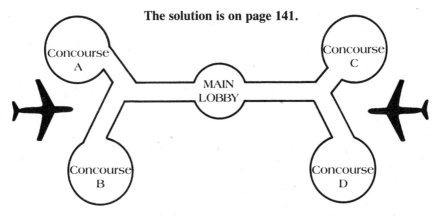

time	passengers	airline	flight #
_____	_____	_____	_____
_____	_____	_____	_____
_____	_____	_____	_____
_____	_____	_____	_____

26 THE DEEP FREEZE

by Evelyn B. Rosenthal

After record-breaking cold during Christmas vacation, Don and four other students returned to college from their hometowns in different states (one is New York) and compared notes about the frigid temperatures on Christmas Day. Each majors in a different subject; one is English. From the following clues, can you find each student's full name (one surname is Goodman), major, home state, and the temperature in his or her hometown on Christmas Day?

1. The temperatures were five degrees apart.
2. The coldest hometown was in Illinois.
3. The middle temperature was in the hometown of the physics major, who is not Alice.
4. Neither Jackson nor Bob is from the warmest place, where the temperature was 0°.
5. Alice's hometown, which is in Tennessee, had a temperature below zero.
6. The student who went to Iowa had a colder Christmas than both Fry and Hayes, but not as cold as the history major.
7. Both Ed's hometown and that of the student from Michigan had temperatures below zero.
8. Carol was not in the warmest place, and Ed was not in the coldest.
9. Both the math major's hometown and Fry's had temperatures below zero.
10. It was below zero in Hayes's hometown.
11. Ives, the art major, spent Christmas in a town which had a below-zero temperature.

The solution is on page 141.

	Fry	Goodman	Hayes	Ives	Jackson	art	Eng.	hist.	math	phys.	Ill.	Iowa	Mich.	NY	Tenn.	temperature			
Alice																			
Bob																			
Carol																			
Don																			
Ed																			
temperature																			
Ill.																			
Iowa																			
Mich.																			
NY																			
Tenn.																			
art																			
Eng.																			
hist.																			
math																			
phys.																			

27 SOUTH OF THE BORDER

by Haydon Calhoun

Last summer, four married couples vacationed in Mexico, traveling separately by various means of transportation. Each family had chosen to vacation there for a different reason: one for the sightseeing, another because it was inexpensive, a third because it was close to home, and the fourth because they had friends living there. Supplied with this information and the following clues, try to determine the name of the couple living in Mexico and, for each of the four tourist couples, their home state, how they traveled to Mexico, their reason for vacationing there, and the number and sex(es) of their children (if any).

1. The Cabots had no friends in Mexico when they arrived, but they met the couple from New Mexico and the Moores, one of whom was visiting the Mexican couple. The Mexican couple took care of the three couples' total of four children—a girl and three boys—while the six tourist parents went out to dinner in the van in which one parent couple had driven to Mexico.

2. One of the daughterless couples knew the Allens before going to Mexico.

3. The California and Texas couples have one child each.

4. The Nobles and the couple who went to Mexico because it was close to home flew there.

5. Neither the Tylers, who are not the couple from Arizona, nor the daughterless couple with the van knew anyone in Mexico before going there.

6. The California couple went to Mexico by train.

7. The childless tourist couple were not concerned about expenses.

The solution is on page 142.

tourist	from	traveled by	reason	number and sex of children

_____, resident

28 MERRY-GO-ROUND RIDERS

by W. R. Pinkston

Mrs. Keyes took her son and five of his new friends from kindergarten to a small amusement park. The children were treated to two consecutive rides on a miniature merry-go-round that had only six different wooden animals, which were numbered. The mounts were placed so that #1 was just in front of #2, which was followed by #3, etc. No child rode the same mount twice. From the clues below, can you determine each child's first name (one is Greg), surname (one is Smith), the mount each rode each time (one was a bear), and each animal's number?

1. Sue and the Cain girl were located directly opposite one another for both rides. They were the only two who exchanged mounts.

2. Mike's second mount was directly opposite mount #2; he was just behind Wilma.

3. On the first ride, Sue's number was lower than Mary's, and Mary was just in front of Mills; Mary moved back one position to the zebra for her second ride.

4. For the first ride, Ted was just ahead of Brown, who was just in front of the llama.

5. The lion's number was higher than the burro's, but lower than the horse's. Mead rode none of these three mounts, but Mary rode one.

The solution is on page 142.

FIRST RIDE

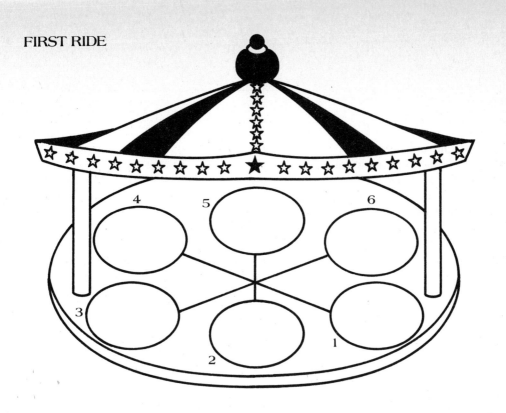

SECOND RIDE

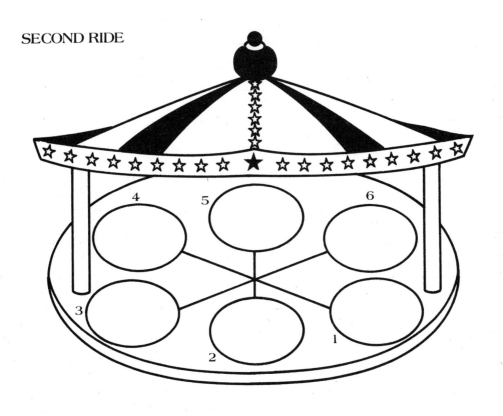

29 KEEPING UP WITH THE JONESES

by S. A. Hilborn

Chris and the other three Jones children have all left home to pursue various careers (one became an interior designer), but they get together as often as they can for visits. From the following clues, which describe Mrs. Jones *and* her children, can you list the four offspring in order of age, find the sex of each, and determine where Mrs. Jones and each of the children lives (one of the five lives in San Mateo) and what each does for a living?

1. The woman who lives in Sacramento went to college in San Diego as did a younger sibling who finally settled there.

2. Mrs. Jones's second-born has always loved horticulture and now works in that field; his older sister pursued another line of work.

3. Rusty lives in Pine Hill.

4. The management consultant is older than Fran and two years younger than the one who lives in San Francisco.

5. Sandy often telephones his brother and sisters.

6. The man in advertising recently flew to San Francisco to visit Lee, who is not the one who owns a hotel.

7. Only the youngest stayed in the city where he had gone to school.

The solution is on page 143.

male or female	name	occupation	city

by Ellen K. Rodehorst

At a certain university, Karen and four other students live on the fourth floor of Stadler Hall, a co-ed dormitory. Three have separate rooms, and two are roommates. One of the rooms is number 409. All of the students are majoring in different fields of study, one of which is psychology. All are from different states; one is from Georgia. From the following information, can you discover each student's full name (one surname is Crane), home state, major, and room number?

1. Three of the students—Peter, the one from California, and the chemistry major, who's not from Pennsylvania—often sit together in the dining hall.

2. Young, who lives in room 402, doesn't date.

3. Bill is not the man who comes from Maine, and neither of them is Mr. Pierce, who is not the history major.

4. Susan, who isn't the one whose last name is Reilly, has dated the student from California and the biology major, but not the man from Wyoming.

5. The math major, who lives alone in room 436, doesn't know anyone in the dormitory, not even Horne, who lives in room 435.

6. Jack, who isn't a biology major, goes skiing every weekend with the woman from Pennsylvania.

7. The dormitory is designed with two wings: room numbers 01 to 25 on each floor are reserved for women; the others are assigned to men.

The solution is on page 143.

room #	student	major	home
_____	_____ _____	_____	_____
_____	_____ _____	_____	_____
_____	_____ _____	_____	_____
_____	_____ _____	_____	_____
_____	_____ _____	_____	_____

31 BUNKMATES

by Diane C. Baldwin

Dave and five other boys share three double bunks placed side by side in a cabin at Camp Wilderness. As it happens, each bunk sports a different color blanket, including a blue one. From the clues below, can you identify each boy by full name (one last name is Lake), home state (including a boy from New York), blanket color, and bunk location? NOTE: All locations are as viewed upon entering the room.

1. Bob, the West boy, and the boy from New Jersey are each in a different set of bunks.

2. The boy from Massachusetts and the Greer boy share the bunk to the right of the one shared by Ted and the boy with the brown blanket.

3. Neither Tim nor the boy with the green blanket have a bottom bunk.

4. One of the two boys from Pennsylvania has a bottom bunk.

5. Bill's bunk has the Banks boy's to its left and the bunk with the red blanket to its right—all three on the bottom.

6. The Byrd boy and Paul share the bunk to the right of the one shared by West on the bottom and the Connecticut boy on the top.

7. A Pennsylvania boy's tan blanket covers the top bunk above the New Jersey boy, who isn't the Winston child.

8. Ted doesn't have a top bunk.

9. One double bunk has a green blanket and a gray blanket.

The solution is on page 143.

_____

_____

_____

_____

_____

_____

_____

_____

32 SUMMER ROMANCES

by Margaret Shoop

Each of six teen-age counselors at a summer camp fell in love with exactly one of the other five and was, in turn, the object of the affections of exactly one of the other five. From the clues below, can you determine with whom each counselor fell in love?

1. Betty fell in love with the boy who fell in love with the girl who fell in love with Mike.

2. Joan fell in love with the boy who fell in love with Sue.

3. Betty didn't fall in love with Jason.

4. Bill fell in love with the girl who fell in love with the boy who fell in love with Joan.

The solution is on page 144.

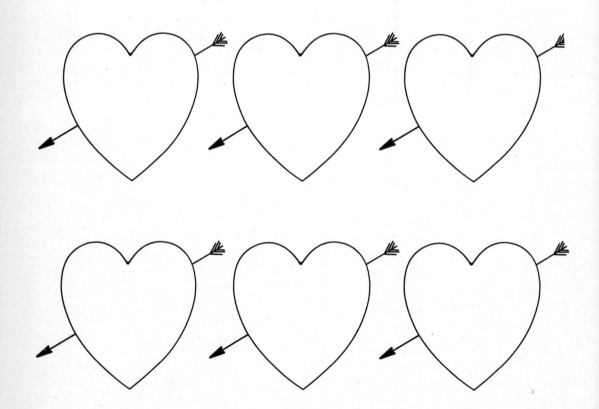

33 BUSINESS PARTNERS

by Frank A. Bauckman

Five married couples, including the Carneys, live on the same Smalltown street. Four of the men operate two two-man business partnerships and the fifth has his own company. From the clues below, try to determine the full names of each of the couples (one of the men is Ivan) and which men are involved in the partnerships.

1. Harry's partner's wife is not Paula; Susan's husband's partner is not Mr. Davis.

2. Mr. Baker's partner is not John or Kevin, nor is his partner married to Marie or Norma.

3. The five women are as follows: Norma, George's wife, Mrs. Abbot, Mr. Evans' partner's wife, and Marie.

4. Norma's husband has a business partner.

5. Neither John nor Kevin is married to Marie or Norma.

6. George's wife is not Paula; Rose is not Mrs. Abbot; John is not Susan's husband.

7. Mr. Abbot has a partner.

The solution is on page 144.

wife	husband	last name	partner

BUSH
BOGGLE

by Mary Marks Cezus

All five couples who live on the east side of the 800 block of Appletree Lane purchased and planted a different kind of flowering bush for their yards recently. Using the diagram on the next page and the clues below, can you determine each couple's first (one of the men is Wayne) and last names, type of bush, and house number?

1. The lilac (which was not purchased by Steve and his wife) was planted at one end of the block; the Pringles live at the other.

2. Vince's next-door neighbors are the Olcotts and the couple who planted the rhododendron.

3. Carol lives on one side of Todd, while the forsythia was planted on the other side of Todd.

4. Edith's house number is less than that of Quigley (who did not plant the mock orange).

5. McDonald (who is not Vince) has Betty as one of his next-door neighbors.

6. The next-door neighbors of the couple who planted the azalea are the Quigleys and Doris and her husband.

7. There are two houses between Zack's house and Alice's house, which is next to the house where the forsythia was planted.

8. There is one house between the Newcombs and the house where the azalea was planted.

9. The Quigleys live next to the house where the forsythia was planted.

The solution is on page 145.

35 DENTIST APPOINTMENTS

by Mary A. Powell

It was time to see the dentist again, so Helen and her whole family—her two daughters, and her two granddaughters—all arranged appointments for the same day. The women drove together to a large shopping center complex where the dentists' offices were located. With so many stores and restaurants within easy walking distance of the dentists' offices, the women planned a day of shopping. From the following clues, can you find the time of each appointment, the full name of each woman, the name of the dentist she saw, and the item she bought (one bought a blouse)? (Note: All five women are married and have taken their husbands' surnames.)

1. The dentist appointments were 50 minutes long and began on the hour. The first was at 10 a.m., and the last ended at 2:50 p.m. No two were at the same time.

2. Marie, her mother, and her cousin had lunch at 1 p.m. Ms. Harris had lunch with her mother at 12 noon.

3. Neither Beth nor Ms. Bradford has children. The woman who bought a pair of shoes has two children. The woman who had the 12 noon dentist appointment has one child.

4. Christine, Ms. Miller, and the woman who bought a dress had appointments with Dr. Stevens, who had lunch at 12 noon. Ms. Carter and the woman who bought a skirt saw Dr. Perkins.

5. Valerie's dentist appointment was two hours after Ms. Vaughn's. Valerie didn't buy the umbrella and Ms. Vaughn didn't buy the skirt.

6. Marie's aunt had an 11 a.m. dentist appointment.

The solution is on page 145.

appt.	first name	last name	dentist	purchase

36 PET VISITORS

by Diane C. Baldwin

On a different day once a week, Peg and four other volunteers bring their pets to visit the Wintergrowth Retirement Home. Mr. Murphy and the four other residents on the first floor each have a favorite pet visitor. One is a retriever. From the clues supplied, can you determine each resident's favorite pet friend and its name and owner, as well as give the weekly visiting schedule?

1. A pet named Pedro comes the day after the cat and the day before Holly's pet.

2. Frisky is scheduled to visit five days after Mrs. Swift's favorite.

3. On Mondays Dave brings a man's favorite pet.

4. A day with no pet visitors separates Cathy's pet's visit and Mr. Stuart's favorite which follows hers.

5. The women's favorites include Pip, Wednesday's pet, and the parrot.

6. Mrs. Jacob's favorite visits three days after Miss Hopper's and the day before a man's favorite.

7. Sam's pet's visit is scheduled earlier in the week then Fancy or the monkey.

8. The pet named Buddy visits the day after the rabbit.

9. No pets are scheduled for Sundays.

The solution is on page 146.

The solution is on page 146.

day	owner	pet, kind	resident
___	___	___	___
___	___	___	___
___	___	___	___
___	___	___	___
___	___	___	___

37 SCIENCE ASSIGNMENTS

by Nancy R. Patterson

A high school science teacher assigned each of nine students an oral report on one of the planets and a written report on another topic. Four of the students are Earl, Gavin, Jill, and Mary. Three of the written assignments covered eclipses, geysers, and tides. The students presented their oral reports in the following order: Mercury, Venus, Earth, Mars, Jupiter, Saturn, Uranus, Neptune, and Pluto. From the clues below, can you figure out each student's assignment?

1. The student who spoke about Saturn wrote about comets.

2. The first and last speakers wrote about two topics that began with same letter.

3. Sally wrote about geodes.

4. Erica was the last girl to speak.

5. Each of two girls spoke about a planet whose name began with her own initial.

6. In no case did the student's name, planet's name, and written topic all begin with the same letter.

7. One student's assigned planet and written topic shared an initial.

8. The student whose paper was on volcanoes spoke later than the one who wrote about constellations, but earlier than Tom.

9. Mark isn't the student who wrote about meteorites.

10. The boy who discussed Uranus was assigned a written topic that began with his own initial, and so was one other boy.

11. John spoke right after the girl who wrote about satellites.

The solution is on page 146.

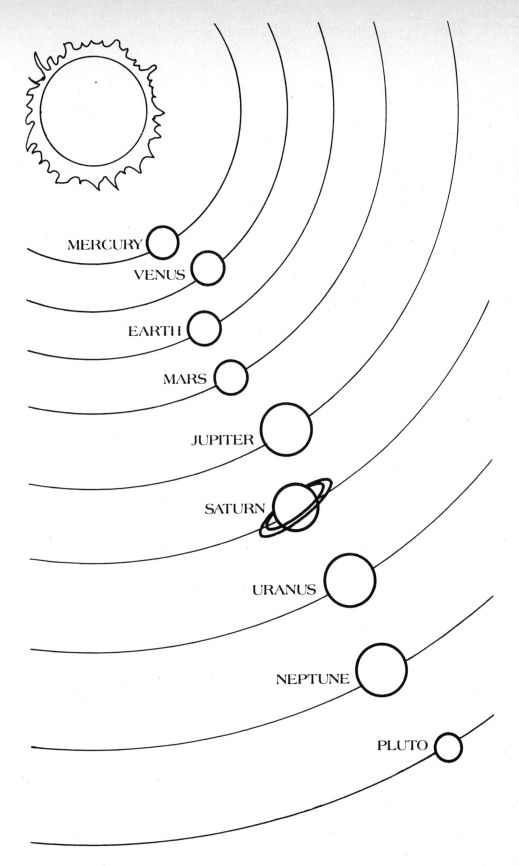

MERCURY

VENUS

EARTH

MARS

JUPITER

SATURN

URANUS

NEPTUNE

PLUTO

SUPER-DUPER SANDWICHES

by Randall L. Whipkey

For lunch last Thursday, five co-workers at the Summerset News Agency went to the Super Duper Sandwich Shop, a restaurant that specializes in unusual sandwiches. Each of the five ordered one of the 162 menu choices—one man chose the "Stomach Stunner"—and each paid a different price. From the clues below, can you determine each man's full name, the sandwich he chose, and how much he paid for it?

1. Earl spent $1.00 more than Owens, who paid twice as much as the man who ordered the "Wide Load."

2. Bill's last name isn't Lemon or Nix.

3. Andy wasn't the one who ordered the "Eiffel Tower."

4. The one who ordered the "Belt Buster" spent $2.00 more than Kane, who spent twice as much as Chad.

5. Dick did not spend the smallest sum.

6. Miller spent twice as much as Andy, who spent twice as much as the man who ordered the "Diamond Jim Brady."

7. The total bill for the sandwiches was $18.00.

8. Neither Andy's nor Chad's last name is Lemon.

9. Miller did not choose the "Wide Load."

The solution is on page 147.

	Andy	Bill	Chad	Dick	Earl	BB	DJ	ET	SS	WL
Kane										
Lemon										
Miller										
Nix										
Owens										
BB										
DJ										
ET										
SS										
WL										

	price	sandwich, person
most	_____	_____
	_____	_____
	_____	_____
	_____	_____
least	_____	_____

39 ROCK COLLECTIONS

by Mary A. Powell

Nearly all of the children in Ms. Simoni's second-grade class have rock collections. When one child asked if she could bring hers in for "show and tell," Ms. Simoni told the class it would be all right to bring up to 12 rocks each, but each collection must be kept in a clear glass jar so they could be seen without being scattered. "That way," she explained, "they won't get lost or mixed up." The next day, four of the children each brought in a dozen rocks for "show and tell." From the following clues, can you find how many of each of the four colors mentioned below each child had?

1. Altogether, 12 white rocks, 12 red rocks, and 12 green rocks were brought in.

2. Each child had at least one of each color; no one had more than 6 of any one color.

3. Cory had more red rocks than green rocks.

4. Lucy had twice as many red rocks as white rocks.

5. Terry, who had the same number of white and green rocks, had more red rocks than anyone else.

6. Jody had three times as many white rocks as black rocks.

7. Lucy had twice as many green rocks as Terry and three times as many green rocks as Cory.

The solution is on page 147.

child	black	green	red	white

HARD LOGIC PROBLEMS

40 FUND-RAISING EVENTS

by Margaret Shoop

On each of the five Saturdays last June, June 2, 9, 16, 23, and 30, a different one of five churches in Center City held a fund-raising event, one of which was a crab picking feast. Each event was put on by a committee headed by a church member. From the clues that follow, can you figure out the date on which each fund-raising event was held, the church sponsoring it (one was St. Luke's Episcopal), and the full name of each church's committee chairperson (one's first name was Mary)?

1. The fourth Saturday's event wasn't the one organized by the committee headed by Forrest.

2. Jean's last name isn't Nichols.

3. Mr. Browder's event was earlier in the month than Joe's, and Joe's event was before the one organized by Nichols.

4. Faith Baptist put on the first Saturday's event.

5. Neither Bill nor Wister headed the committee that put on the barbecued chicken dinner.

6. The fund-raising event of June 30 wasn't the pancakes-and-sausage breakfast.

7. The fish fry was held on the Saturday preceding the event organized by Jean and her committee; Jean's event was held on the Saturday preceding the event put on by the First Unitarian Church.

8. Forrest isn't a Methodist. The event headed by Forrest was held a week later than the one headed by Alan.

9. The event organized by Wister was one week before Trinity United Methodist's fund-raising event, which was two weeks before the spaghetti dinner, which wasn't on June 30.

10. Joe isn't a member of First Unitarian Church.

11. Woods isn't a member of St. Mary's Catholic Church.

The solution is on page 147.

June	chairperson	church	event
2	_____ _____	_____	_____
9	_____ _____	_____	_____
16	_____ _____	_____	_____
23	_____ _____	_____	_____
30	_____ _____	_____	_____

41 WEEKLY CHORES

by Diane Yoko

The six tenants in a small apartment building all used the washing machine last week, one each day Monday through Saturday; each also vacuumed twice (on two different days) during the same period. From the following clues, can you deduce each tenant's full name, laundry day, and vacuuming days?

1. These four women washed clothes on consecutive days: Wanda, Ms. Hogan, Flora, and one of the women who vacuumed on Monday.

2. Vicki, who did not vacuum on Wednesday, did her laundry earlier in the week than Ms. Magers.

3. Ms. Gleason did not vacuum on Thursday.

4. Ms. Bookman did her laundry the day after one of the women who vacuumed on Friday.

5. These four women washed clothes on consecutive days: Missy, a woman who vacuumed on Saturday, Ms. Lange, and Laura.

6. No woman vacuumed on consecutive days.

7. Susan did her laundry the day before Ms. Bates, who washed her clothes the day before one of the women who vacuumed on Monday.

8. Each day from Monday through Saturday, either Ms. Bates or Ms. Lange did laundry or vacuumed.

9. Each day from Monday through Saturday, either the woman who washed clothes on Monday or the one who washed clothes Saturday performed one of the chores.

10. At least two women vacuumed on Tuesday.

The solution is on page 148.

LAUNDRY

Mon.	Tues.	Wed.	Thurs.	Fri.	Sat.

VACUUMING

Mon.	Tues.	Wed.	Thurs.	Fri.	Sat.

42 BABY-SITTERS

by Randall L. Whipkey

Six mothers who live on Summerset Street have formed a baby-sitting co-op, in which each woman cares for the child of another one evening each week. Last week, beginning Monday, each woman baby-sat on a different evening; none did so on Sunday, and each left her child with another mother one evening during the week. From the clues that follow, can you determine each mother's full name, the full name of the child with whom she baby-sat (one boy is Ian), and on which evening?

1. Helen baby-sat the night before Mrs. Wilson, who baby-sat the night before the woman who watched the Timmons boy. The Timmonses did not have a baby-sitter on Saturday.

2. Mrs. Usher baby-sat the night before Gloria, who baby-sat the night before the woman who watched Linda.

3. Doris baby-sat earlier in the week than Mrs. Rivers, who baby-sat earlier in the week than the woman (who is not Mrs. Short) who baby-sat Marty.

4. Mrs. Wilson did not baby-sit the Rivers child.

5. Helen baby-sat with Nancy.

6. Ellen did not baby-sit with Julie.

7. Mrs. Usher baby-sat with one of the girls.

8. The woman who kept the Vance girl baby-sat earlier in the week than Carla.

9. Mrs. Timmons baby-sat earlier in the week than Kevin's baby-sitter, who did not baby-sit him on Wednesday.

10. Faith, who is not Mrs. Usher, baby-sat earlier in the week than the woman who kept the Usher girl.

11. Marty's last name isn't Wilson.

The solution is on page 149.

	baby-sitter		baby-sat	
Mon.				
Tues.				
Wed.				
Thur.				
Fri.				
Sat.				

43 HOW OLD IS YOUR CAT?

by Mary A. Powell

The four Thompson brothers and their families live in the same town and will use any excuse to get all of the families together. Each Thompson family has two children and a cat, but the cats are usually left at home during family gatherings. From the following clues, can you find the names and ages of the cat and the children (including Brad and a 7-year-old) in each of the families? (Note: All ages are in whole numbers.)

1. Ted is twice as old as Tinker Bell (a cat), who is twice as old as Tanya.

2. Julie and the cat named Fuzzy are the same age; they are twice as old as Julie's cat.

3. Nadine is four years older than her sister, who is four years older than Ted's brother.

4. Sam's sister is the same age as the cat named Pixie.

5. Spooky (a cat) is two years older than Rodney and two years younger than Rodney's brother.

6. Carrie's brother is ten; Sam is the same age as Carrie's cat.

7. The oldest child is twelve.

8. In one family, the cat is older than both children; in each of the other three families, the children are older than the family's cat.

The solution is on page 149.

The solution is on page 149.

child (1)	age	child (2)	age	cat	age

44 DIET CHALLENGE

by Ellen K. Rodehorst

Recently, John and three fellow workers challenged one another to diet for a month. Each of the employees has a different job; one is a sales representative. Each person is a different age (only one age is an odd number), but all are in their thirties. At the start of the month, the lightest person weighed 150 pounds. By the end of the month, all reported weight losses. From the data below, can you determine each dieter's name, age, job, weight loss, and final weight?

1. The oldest dieter, who's not the typist, is three years older than the dieter who initially weighed 180 pounds, who is not Diane.

2. The thirty-two-year-old is not the youngest dieter.

3. Diane cheated a little on her diet but still lost fourteen pounds, while the clerk, who's not Mary, lost only twelve pounds; still another dieter lost sixteen pounds.

4. At the end of the diet, the accountant weighed one pound less than Alex and was the most successful dieter, with a 10 percent weight loss.

5. The youngest dieter, who weighed 170 pounds before dieting, is seven years younger than the dieter who lost only five pounds.

The solution is on page 149.

age	name	job	lbs. lost	final weight

45 LOST IS FOUND

by Mary A. Powell

When Mrs. King found an unmailed letter under her car seat, she attached a cover letter and mailed it on to her sister. This started a chain of events in which four other members of Mrs. King's family also found and returned an item belonging to another family member. From this information and the clues, can you find the full name of each person, the item each found, the relationship and name of each person to whom the item was sent, and the order in which each was sent? (Note: Unmarried children have the same last names as their parents.)

1. Mrs. King is the only person who sent an item before receiving one.

2. Joyce and Mrs. Castle each sent an item to a daughter; one sent photos of a family picnic; the other received a knitted sweater pattern.

3. The Bishop twins, who attend different colleges, are not married; neither they nor Paula have children. Helen has no sisters.

4. Christine, who didn't send the lost sunglasses, sent an item after Karen received one.

5. The person who received a lost textbook from her sister sent something to her grandmother.

6. Helen and Joyce had no correspondence with each other.

The solution is on page 150.

Mrs. _____ King
item: _____
to: _____ _____
rel.: _____ sister _____

sender: _____
item: _____
to: _____
rel.: _____

sender: _____
item: _____
to: _____
rel.: _____

sender: _____
item: _____
to: _____
rel.: _____

sender: _____
item: _____
to: _____
rel.: _____

82

46 MR. TAXI'S FARES

by Diane Yoko

Mr. Taxi had a busy afternoon recently, when he picked up and drove five people to their destinations. From the following clues, try to deduce his fares' full names, their pickup locations and destinations, and the cost of each ride.

1. Four of the rides were Ms. Becker's, the one starting from Briar Lane, the one to the zoo, and the one to the theater; Alice is not Ms. Becker.

2. The ride to the mall was ten dollars less than Alice's.

3. Neither Ms. Parker nor Sam was picked up on Briar Lane.

4. The one who was picked up on Beacon Drive did not go to the zoo.

5. Haley did not go to the theater.

6. Smith, who is not Eileen, was not the one picked up on Lexington.

7. The ride from Beacon Drive cost twice as much as the one to the bank, which was four dollars less than Becker's.

8. Mahoney was picked up on Broadway.

9. Debra's ride cost half as much as the one to the airport and five dollars less than Mahoney's.

10. Eileen was not the one picked up on Maine Street.

11. Carl, who did not go to the zoo, was picked up on Lexington Avenue.

The solution is on page 150.

rider	starting point	destination	cost

47 WHO HAD THE DAY OFF?

by Haydon Calhoun

The Family Cafeteria is owned and operated by three married couples, including the Smiths, and one other person. Last week, each of the seven people, who include Sam, was off a different day, Sunday through Saturday. From the following clues, can you determine each person's full name and day off? (Note: All the wives have adopted their husbands' surnames.)

1. Will was off before at least four of the others.

2. Mr. Marsh was off before at least two others, but not before Tina.

3. Molly's day off preceded her husband's day off; there was one day between them. The same was true of Mrs. White and her husband.

4. Fay wasn't off Friday or Saturday.

5. Ted was off before at least one woman.

6. Sid's wife isn't Molly or Tina.

7. Mr. Flint, who isn't married to Molly, was off after Sid.

8. Mr. White wasn't off Tuesday.

The solution is on page 151.

DAY OFF

SUNDAY	MONDAY	TUESDAY	WEDNESDAY	THURSDAY	FRIDAY	SATURDAY

48 NAME MATCH

by Margaret Shoop

Ms. Adams, Ms. Saunders, and three other players on a women's softball team are known by their first and middle names together, rather than by just their first names; one is known as Ida Lee, and one name of another is Lou. One of the five plays third base. From the clues that follow, can you determine each woman's full name and team position?

1. No two of the five middle names begin with the same letter.

2. The one whose first name is Mary plays shortstop.

3. One's middle name is Sue; she is not Ms. Irving.

4. Marie is one name of the woman who plays second base.

5. The one whose first or middle name is Jane plays first base; her other given name is not Lou.

6. One of Ms. Luce's given names is Sarah.

7. One's first name is Anne, and another's middle name is Anne; one of the two is Ms. Madison, who doesn't play first base, and the other is the catcher.

8. The middle and last names of only two women begin with the same letter.

The solution is on page 151.

1st base	_____	_____	_____
2nd base	_____	_____	_____
3rd base	_____	_____	_____
shortstop	_____	_____	_____
catcher	_____	_____	_____

49 MOWING LAWNS

by Julie Spence

During the summer, Nick Andrews earns money by mowing his neighbors' lawns. Last week, Nick was lucky; it didn't rain all week, so each day, Monday through Friday, Nick mowed a lawn in the morning and went swimming in the afternoon. Nick was paid a different whole number of dollars for each lawn he mowed, the most he was paid was $12, the least $8. From the information below, can you determine the full name of each neighbor whose lawn Nick mowed last week (one first name is Sally, one surname Wilson), and how much Nick earned each morning?

1. Nick mowed a woman's lawn on Wednesday and she paid him more than Mildred but less than at least one of the others.

2. Nick was paid more for the lawn he mowed on Friday than Frank paid him, but Frank paid him more than Norris, who was not the one whose lawn Nick mowed on Thursday.

3. John's lawn was mowed the day before Miller's, and John paid Nick two dollars more than Miller.

4. Joe, who did not pay Nick the most, paid him more than Carpenter.

5. Nick mowed Hilton's lawn on Tuesday.

The solution is on page 152.

Mon.	_____	_____	$_____
Tues.	_____	_____	$_____
Wed.	_____	_____	$_____
Thur.	_____	_____	$_____
Fri.	_____	_____	$_____

50 MILL VALLEY PROGRAM

by Nancy R. Patterson

Hank was one of the performers when twelve children from grades one through six at Mill Valley School presented a program for their parents. Six of the children sang, and the other six acted in various roles (one played the role Cinnamon). Each of the singers wore a different color of robe (one wore white). From the clues below, can you decide each child's grade and role (if an actor) or robe color (if a singer)?

NOTE: You may assume that an older child is in a higher grade.

1. One girl and one boy from each grade participated, one by acting and the other by singing.

2. The six actors are Nick, Ginny, a sixth-grade boy, the second grader who played Pinwheel, and the pupils who played Nutmeg and Carousel.

3. Carol, who acted, is older than the pupil who played Pippin.

4. Paul isn't in Kate's grade.

5. Debby isn't the oldest girl, but she's older than Sandie.

6. The actor who played Clove is in the same grade as the singer who wore lavender, and Mark is in the same grade as the singer who wore blue.

7. The three boys who sang are John, a fourth grader, and the one who wore green; the three girls who sang are Sandie, a third grader, and the one who wore pink.

8. Lisa is younger than Ginny but older than Bob.

9. The actors whose roles had spice names are all of the same sex.

10. The singer who wore yellow is in the next grade above the one who wore blue.

The solution is on page 152.

ACTORS

grade	child	role

SINGERS

grade	child	robe

51 KEEPING THE LIBRARY OPEN

by Evelyn B. Rosenthal

With many legal holidays now falling on Monday, our community faced the prospect of the local branch of the library being shut for several three-day weekends. To avoid this, three people volunteered to help keep the library open and the three librarians agreed. The last long weekend, each volunteer worked two of the three days, and a different librarian was there to supervise on each of the three days; each librarian then had a free day—Tuesday, Wednesday, or Thursday—the following week, and one of the volunteers acted as a substitute that day. From the following clues, can you find the full name of each of the six (one first name is George, one last name Anders) and which days—Saturday through Thursday—he or she worked?

1. Helen and Ms. Brooks each worked alternate days.

2. Only John had all his working days consecutive.

3. Meg had all her free days consecutive.

4. Drake, who is not Ned, worked Saturday.

5. Ms. Calhoun is not a volunteer; Flint is.

6. Ann did not work Saturday.

7. One of the six is Ms. Eastman.

The solution is on page 153.

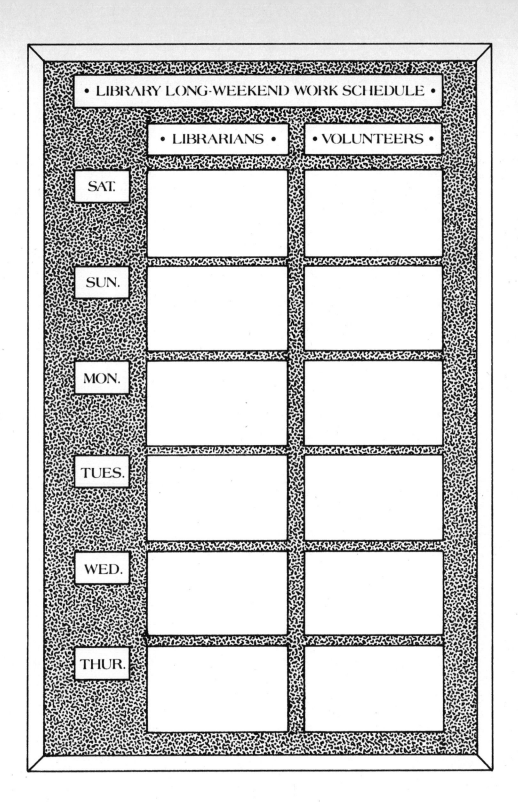

• LIBRARY LONG-WEEKEND WORK SCHEDULE •

	• LIBRARIANS •	• VOLUNTEERS •
SAT.		
SUN.		
MON.		
TUES.		
WED.		
THUR.		

52　DRIVING TESTS

by Nancy R. Patterson

A particularly strict examiner gave only 70 and 80 as scores the day Becky, Dave, and four other high school students took their driving tests. Three of these candidates were examined in the morning and the other three in the afternoon. Each candidate received two separate scores of either 70 or 80, one for a written quiz and the other for a practice drive. To earn a driver's license required a combined score of at least 150. From the following clues, can you deduce each candidate's full name (one surname is Peters), whether he or she was tested in the morning or the afternoon, and his or her scores on the two parts of the test?

1. Of the six, only Vane failed to earn a driver's license.

2. Gary, who is not Simms, is taking a driver-education course.

3. Jenny, whose combined score was higher than either of the other girls', took her test at the same time of day as the White girl.

4. No boy received the same score on both parts of the test as any other boy.

5. Neither Matt nor Jenny is Simms.

6. The Taylor girl did better on the quiz than the drive, but that was not true of the Root boy.

7. On each test, the total of morning scores exceeded the total of afternoon scores; the morning candidates received altogether two more scores of 80 than the afternoon candidates.

8. Heidi took her test in the afternoon.

9. Everyone enrolled in a driver-education course earned a license.

The solution is on page 153.

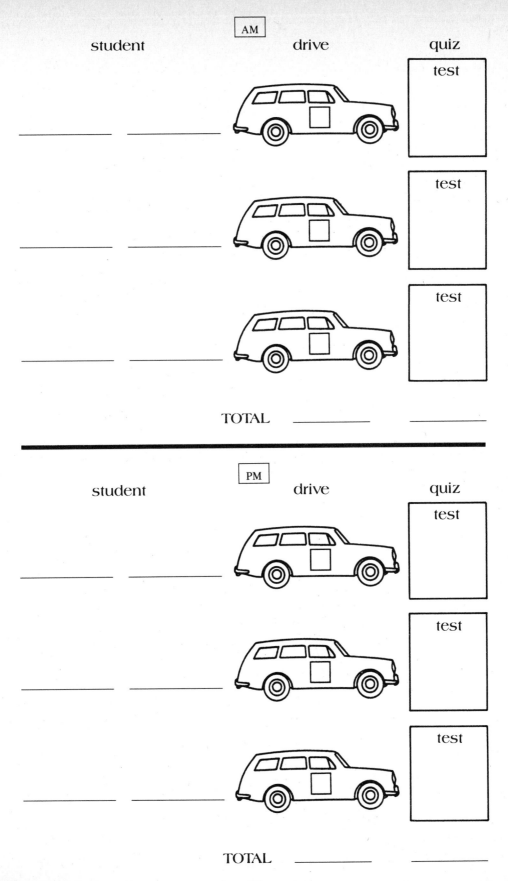

53 THE SHOPPERS

Ms. Newton and five other women from the same neighborhood all went shopping one morning at a local department store. Each woman went directly to the floor carrying the article that she wanted to purchase and shopped on no other floor; each made exactly one purchase; and each shopped on a different one of the six floors. From the clues which follow, can you determine the floor on which each of the six shopped and the article that she purchased?

1. One of the other shoppers saw Ms. Vail on the down escalator between the fourth and fifth floors.

2. Ms. Stern and the woman who bought the tablecloth got on the elevator together when they entered the store; the latter got off before Ms. Stern.

3. Ms. Patrick's purchase was made on the third floor.

4. Ms. Weeks bought a jacket for her husband.

5. One of the six purchased a crib in the infants' department, which is three floors below the department where the lamp was purchased.

6. The washing machine was purchased three floors above the men's department.

7. The new dress was not purchased on the first or fifth floor.

8. Ms. Thomas didn't buy the crib and didn't make her purchase above the third floor.

The solution is on page 154.

floor	shopper	purchase
6	_____	_____
5	_____	_____
4	_____	_____
3	_____	_____
2	_____	_____
1	_____	_____

92

54 SPRING BULB ORDERS

by Diane Baldwin

When David sold spring bulbs for a scout project, the Clevelands and four other families placed orders with him. Each family ordered a different variety, between two and five dozen, and planned a different place for planting, including a spot by a front porch. From the clues supplied, can you identify each family's order by size, bulb variety, and planting place, as well as tell the order in which David took the orders? (Note: All orders are in one dozen units.)

1. The crocus order was half as large as the third order.

2. The Crocketts placed their order just after a smaller order of hyacinths and just before the bulbs to be planted by a garage.

3. The five orders added up like this: the second customer's order and the daffodil order and the Morrises's order combined equaled the number of dozens to be planted by a front walk plus the only three-dozen order.

4. The largest order, which wasn't the one for jonquils, was placed just before the bulbs to be planted around a mailbox.

5. The Stouts' order was larger by one dozen than the order immediately following theirs, which was to be planted around a birdbath.

6. Tulips weren't the fourth order, nor were they ordered by the Merchants.

The solution is on page 155.

The solution is on page 155.

	last name	amount	flower	place
1				
2				
3				
4				
5				

55 ALIEN LUV

by Julie Spence

The good ship Puttinuon was on a twenty-five-year space mission to explore the universe. Therefore, it came as no surprise when, during the third year of space travel, Perfec and four other male crew members and five female crew members asked their ship's captain to marry them. The captain set aside a day of feasting and celebration and joined the five couples in marriage in five consecutive ceremonies. The last names of these crew members are, in alphabetical order: Aleon, Frendle (who is a male), Imuss, Yano, Ntel, Ophens, 4ee, Ther, Uwood, and Withme. From the information below, can you determine the full names of each couple who were joined in marriage (one first name is Pherst), what planet each hails from (one female crew member hails from Telme, and another crew member is from Ovkorsnot), and in what order the five ceremonies took place?

1. The crew member who hails from Butduu and her betrothed were married just before Goowt and his betrothed and just after Hy, who is not the female who hails from Lykta.

2. Uwood's wedding, which was not the first, took place just before Iaman and his betrothed were married, which was before the female who hails from Weir was wed.

3. Just before his own ceremony, Ntel stood witness to the wedding of Ru and his betrothed.

4. The female who hails from Conphes, who is not Watta, was the fourth to be married.

5. Withme, who hails from Idluvit, was not the first or last to be wed.

6. Kis, who is not the one who hails from Sheram, and his betrothed were married after at least one other couple and just before Mab, who was married just before the crew member from Som'i.

7. The crew member from Chother, who did not stand witness to any ceremony, was married just after Imuss.

8. Nouh's wedding was not the first, but it took place before Aleon and his betrothed were married, which was before Yano was wed.

9. Ru is not the male whose last name is 4ee.

10. Hy's last name is not Ophens.

The solution is on page 155.

	wives			husbands		
order	first name	last name	planet	first name	last name	planet
⎯⎯	⎯⎯⎯	⎯⎯⎯	⎯⎯⎯	⎯⎯⎯	⎯⎯⎯	⎯⎯⎯
⎯⎯	⎯⎯⎯	⎯⎯⎯	⎯⎯⎯	⎯⎯⎯	⎯⎯⎯	⎯⎯⎯
⎯⎯	⎯⎯⎯	⎯⎯⎯	⎯⎯⎯	⎯⎯⎯	⎯⎯⎯	⎯⎯⎯
⎯⎯	⎯⎯⎯	⎯⎯⎯	⎯⎯⎯	⎯⎯⎯	⎯⎯⎯	⎯⎯⎯
⎯⎯	⎯⎯⎯	⎯⎯⎯	⎯⎯⎯	⎯⎯⎯	⎯⎯⎯	⎯⎯⎯

We found it helpful to determine male versus female names and to list them accordingly. This space is provided for that purpose.

EDUCATIONAL OPPORTUNITIES

by Virginia C. McCarthy

Mr. and Mrs. Lerner, who have five school-age youngsters, firmly believe that children should be given the opportunity to learn as much as possible during their formative years. As a result, Nanette and the other Lerner children are now each taking both a weekly music lesson (one is learning piano) and a weekly class (one is in drama). The lessons and classes take place on various days of the week, Monday through Saturday; none of the children has both activities on the same day of the week. From the following clues, can you find each child's musical pursuit and class and day of the week for each?

1. The guitar lesson is earlier in the week than Jonathan's class and later in the week than Martha's music lesson.

2. Karen does not take violin lessons.

3. Only the pottery class and the flute lesson take place on Tuesday.

4. The child who takes guitar lessons also attends a woodworking class.

5. Saturday's agenda includes classes for two of the girls and a music lesson for one of the boys.

6. The child who takes clarinet lessons attends a photography class on Wednesdays.

7. Of the music lessons, only Leonard's is on Monday and only the violin student's on Thursday.

8. The drawing class takes place on Friday.

The solution is on page 156.

The solution is on page 156.

MONDAY	TUESDAY	WEDNESDAY	THURSDAY	FRIDAY	SATURDAY

57 BOBBING FOR APPLES

by Julie Spence

When Jamie Smith gave a Halloween party, he discovered that the Anderson child, the Vail child, and three other party guests particularly liked bobbing for apples. From the clues below, can you determine the full names of these five guests (one is Kari), how many apples each got in a two-minute period, and what Halloween costume each was wearing at the party?

1. Eric got twice as many apples as the child dressed as a black cat, and the Wick child got twice as many apples as the one dressed as a hobo. (Note: Four of the children are mentioned in this clue.)

2. Susan wasn't dressed as a hobo.

3. Jane wasn't dressed as a vampire.

4. Adam, who isn't the Strom child, got fewer apples than the Blake girl.

5. The one in the ghost costume got six apples.

6. The one dressed as a witch got more apples than Susan did but fewer than at least one other.

7. The two dressed as a hobo and a black cat did not get the same number of apples.

8. Unfortunately, the Strom child didn't get any apples.

9. None of the five children has the same first and last initials.

The solution is on page 156.

child		costume	# of apples
_____	_____	_____	_____
_____	_____	_____	_____
_____	_____	_____	_____
_____	_____	_____	_____
_____	_____	_____	_____

58 FORGETFUL FOOTES

by Diane C. Baldwin

The forgetful Foote family fairly flew from their flat up Fleet Street to the freeway for a fling in Florida. Before they passed five intersections, Felix and the other four Footes found they each had forgotten something (including the food) and were forced back to their flat to fetch it. Each turnaround was made at a different street intersection, one of them at Field Street. From the following clues, can you figure out who forgot what and in what order, and find the order of the intersections on Fleet Street from the Foote's flat to the freeway?

1. The second turnaround began at the street following Flag Street and the street before Fred had to return to the flat.

2. The men were responsible for the return for the film, the turnaround at Fig Street, and the fifth trip back.

3. Fork Street followed the one where they returned to fetch the flashlight and preceded the one where a woman had them make their first turnaround.

4. The final trip back didn't begin at Frond Street, nor was it the one to fetch the fan.

5. Frank's forgetfulness turned them back one block and one return trip following Francine's.

6. One woman was the third to forget, while the other woman turned them back at Flag Street.

7. They returned for the fiddle just before the trip back that began at Frond Street and just following the one requested by Flo.

The solution is on page 157.

ORDER OF STREETS

FLEET STREET

ORDER OF TURNAROUNDS

who	where	what
1. _____	_____	_____
2. _____	_____	_____
3. _____	_____	_____
4. _____	_____	_____
5. _____	_____	_____

59 WHO SAW THE GOAT?

by Mary A. Powell

Each year, the Jackson family takes a vacation trip by station wagon, so young Jonathan and each of the other three children have a window seat. This year, the children devised some new games that involved counting objects seen from the windows. They counted trees, cars, houses, animals, etc. The rules to "Counting Animals" were: Two children on opposite sides of the car counted animals seen from his or her side window for three minutes, then the other two (also on opposite sides of the car) counted animals for the next three minutes. The non-counters made certain the counters didn't cheat. The winner was the one who counted the most during the six-minute game. During their first game in Wyoming last summer, each child spotted from 1 to 9 horses, 1 to 9 cows, and 1 to 9 sheep, plus one other type of animal not seen by any other child. From the following clues, can you find how many of each type of animal each child spotted during that game?

1. No two spotted the same number of any one type of animal; no two saw the same total number of animals.

2. Jennifer saw twice as many horses as sheep.

3. Julie saw three times as many sheep as horses.

4. Jeremy spotted three llamas and only one horse.

5. Julie saw one more cow than Jeremy.

6. The child who spotted the jack rabbit had the highest grand total; the child who saw the goat had the lowest total.

7. One child, who saw the single deer, saw the most sheep; another child who saw the fewest cows also saw the most horses; Julie, a third child, saw the most cows.

8. Twelve horses and twelve cows were spotted altogether.

The solution is on page 157.

child	sheep	horses	cows	other	total

60 SUBSTITUTE TEACHERS

by Diane C. Baldwin

One recent Monday-through-Friday school week, Ms. Foley and four other teachers at Chatham High needed to use substitutes when they were absent, each for a different number of days. Each teacher taught a different subject, including art, and was replaced by a different substitute (one was Ms. Block). Using the information below, can you determine each teacher's subject, substitute, and days absent?

1. The history teacher was present on Wednesday.

2. Ms. Windsor filled in for the woman who teaches math, but not on Monday or Tuesday.

3. The Latin teacher and Mr. Nelson and the teacher whose sub is Mr. Kerwin and the teacher who was absent all week were the only teachers absent Tuesday.

4. Mr. Taylor substituted for one of the women.

5. Only one teacher, except, of course, for the man absent only one day, was not absent on all consecutive days.

6. Mr. Taylor and Ms. Windsor were the only substitutes in on Wednesday and neither replaced Ms. Morgan or the teacher out for two days.

7. The English teacher was absent one more day than Ms. Davis.

8. The woman who replaced Mr. Lynch, though not on Wednesday, didn't fill in as often as Ms. Clark.

The solution is on page 158.

teacher	subject	substitute	day(s)

61 PLAYGROUND DUTIES

by Mary A. Powell

Ms. Johnson, the principal at Zionsburg Elementary School, had the task of making out the playground duty schedule for the first two weeks of school. There are seven teachers at the school, one for each grade (K–6), and five duty times each day (Monday through Friday). All teachers were available for morning bus duty and afternoon playground duty, but only primary (K–3) teachers could take morning recess duty and lunchroom duty, and only upper-grade (4–6) teachers could take the noon playground duty. As shown on the opposite page, in the first draft, Ms. Johnson rotated all teachers alphabetically through morning bus duty, rotated primary teachers through the recess duty list by grade level (starting with the kindergarten teacher), continued with those names through the lunchroom duty list, then rotated the upper-grade teachers through the noon duty list by grade level (starting with the fourth-grade teacher). On the final draft, she made adjustments where necessary, then filled out the afternoon duty list to even out the duties, thus producing a schedule that was as fair as she could make it. From this information and the clues, can you find the name and grade level of each teacher, and the final duty roster which Ms. Johnson posted?

1. No teacher had more than one duty per day. Each teacher had at least one duty-free day per week. Only the third-grade teacher had fewer than three duty-free days in the two-week period.

2. To eliminate conflicts, where a teacher was scheduled for two duties in one day on the first draft, Ms. Johnson made only four switches: Baker and Evans were switched on both a recess and a lunchroom duty the first week, Anderson and Evans were switched for a recess duty the second week, and Carter and Dewey were switched for a noon playground duty the second week.

3. Ferguson had two recess duties and two duty-free days the first week.

4. Carter had Monday afternoon duty the second week; Grayson had Thursday afternoon duty the second week.

5. In the final draft, Baker had no Monday duties the first week; Anderson had no Friday duties the first week; Dewey had no Tuesday duties the second week.

The solution is on page 158.

MS. JOHNSON'S FIRST DRAFT

1st week

	BUS (all tchrs.)		RECESS (K-3)		LUNCH (K-3)		NOON (4-6)		AFTERNOON (all tchrs.)	
	gr.	teacher	gr.	teacher	gr.	teacher	gr.	teacher	gr.	teacher
Mon.		Anderson	K		2		4			
Tues.		Baker	1		3		5			
Wed.		Carter	2		K		6			
Thurs.		Dewey	3		1		4			
Fri.		Evans	K		2		5			

2nd week

Mon.		Ferguson	1		3		6			
Tues.		Grayson	2		K		4			
Wed.		Anderson	3		1		5			
Thurs.		Baker	K		2		6			
Fri.		Carter	1		3		4			

62 TREASURE ISLAND

by Evelyn B. Rosenthal

Flint and four other pirates landed on Treasure Island with the crews of their ships, and each buried a chest, no two of which held the same contents. The five sites are shown on the map; in this instance not X's but A, B, C, D, and E mark the spots. From the following clues, can you find the site chosen by each pirate, the name of his ship (one was the *Silver Snail*), and the contents of his chest?

1. Pegleg's chest was farther from the *Golden Gull*'s than from the chest of coins, and farther from the coins than from the *Silver Skull*'s.

2. Bones, who trusted no one, had removed his treasure before reaching Treasure Island and filled his chest with sand.

3. Blackbeard's chest was as far from the chest of gems as the chest of ingots was from the *Golden Gleam*'s chest, which was not Blackbeard's ship.

4. The *Golden Goat*'s chest was directly west of the chest of gems.

5. Redbeard chose the easternmost site.

6. Bones' chest was directly south of the chest of coins.

7. The *Golden Goat*'s chest was nearer to the chest of arms than it was to the chest that held sand.

The solution is on page 159.

A _____ _____ _____

B _____ _____ _____

C _____ _____ _____

D _____ _____ _____

E _____ _____ _____

Treasure Map

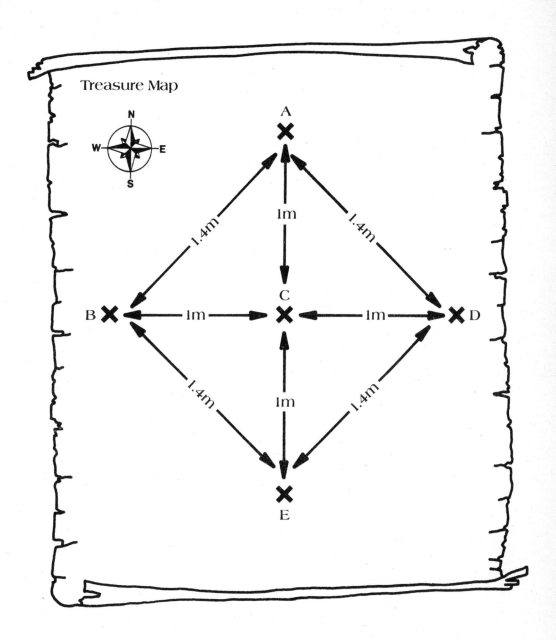

63 PETE'S PAINT SHOP

by Mary A. Powell

When Pete's Auto Paint Shop opened with the slogan, "Paint Your Car In Half A Day," his first week's schedule was filled (normally it takes longer than half a day to sand and paint a car, but Pete installed some sophisticated new equipment that permitted it). One car was scheduled for each morning and afternoon, Monday through Friday. From the following clues, can you reconstruct Pete's schedule with the full names of the ten owners (one first name is Christine), the original color of each car, the color each was painted, and the color of each car's interior (one interior is white)?

1. Linda's car was painted in the morning the day after Mr. Peterson's and the day before the blue car was painted white.

2. Nick's car was painted the day before Ms. Quincy's and the day after Baker's.

3. Brad's car was painted two days before Miller's; both were painted the same color, which was not red or white.

4. Janet's car was painted before the Owens car; both have the same color interior and both were painted in the morning.

5. Donna's car was originally the same color as its interior; it was painted the same day the green car was painted another color.

6. Phil's car, which does not have a black interior, was originally tan; it was painted in the morning the day before the Lowell car.

7. One day's work order was for Nixon's car and the car that was originally yellow; another day's work order was for Michelle's car and the car that was originally white.

8. David and Warner both had their cars painted yellow; both were painted in the morning.

9. Two of the cars have brown interiors; the exterior of one of these two cars was brown originally, the other was painted brown.

10. Pete saw no men on Tuesday and no women on Thursday. He saw no yellow cars (before or during painting) on Wednesday or Thursday.

11. The red car was painted white two days before the black car was painted red.

12. Both gray cars were painted in the afternoon; one was painted black, the other was painted blue.

13. Ken, Curry, and Dawson all have cars with black interiors; Michelle does not.

14. Brad's car was painted before David's car. Ken's car was painted before Phil's car. Ken is not Peterson.

15. Six of the cars have interior colors the same as their original color; two have interior colors the same as their new exterior color. There are no yellow or green interiors.

16. Only one car was painted brown, only one blue, and only one green. None of the cars was painted the same color as its original color.

17. Michelle is not Quincy.

18. Linda is not Nixon or Dawson.

19. Three of the women had their cars painted in the afternoon.

The solution is on page 160.

MON

AM
customer: _____
orig. color: _____
new color: _____
interior: _____

PM
customer: _____
orig. color: _____
new color: _____
interior: _____

TUES

AM
customer: _____
orig. color: _____
new color: _____
interior: _____

PM
customer: _____
orig. color: _____
new color: _____
interior: _____

WED

AM
customer: _____
orig. color: _____
new color: _____
interior: _____

PM
customer: _____
orig. color: _____
new color: _____
interior: _____

THURS

AM
customer: _____
orig. color: _____
new color: _____
interior: _____

PM
customer: _____
orig. color: _____
new color: _____
interior: _____

FRI

AM
customer: _____
orig. color: _____
new color: _____
interior: _____

PM
customer: _____
orig. color: _____
new color: _____
interior: _____

64 PARADES AND PICNICS

by Cheryl L. McLaughlin

The Plums and five other Oakdale families—each consisting of a mother, father, and two children—celebrated Independence Day by attending the parade, the picnic, or both. One of the fathers is Bill, one of the mothers is Dot, and three of the children are named Andy, Danny, and Carrie. From the following clues, can you determine the full names of all the families and the way(s) each family celebrated?

1. The families who attended the picnic were: the Powells and their daughters, Kathie and Candie; Kent and his wife Eve and their children; and Mark and his wife and children, who do not include Robby.

2. The Pierce family, Eric and his wife and children, and Joey and his parents and sister (who isn't Tracie) were the three families who went to the parade but not the picnic.

3. Ann and her husband, son, and daughter Tracie went to the parade with John and his wife and daughters; Gary and his wife, who isn't May, and their daughters went to the parade with the Pyle family, who include a daughter named Vickie.

4. Kay and her husband and family, who aren't the Pecks, did not attend the parade.

5. Neither Ava and her husband and children nor Bobby and his family attended the picnic; Cindie Peck and her family did not attend the parade.

6. One of May's daughters, Abbie, is the same age as the Payne girl.

The solution is on page 161.

event	wife	husband	last name	children
____	____	____	____	____
____	____	____	____	____
____	____	____	____	____
____	____	____	____	____
____	____	____	____	____

65 WEATHER REPORTS

by Mary A. Powell

One October evening while the Hills were reading the local temperature listings in the paper, they also scanned the national weather reports to find out how their four married daughters were faring that day. From the following clues, can you find which five cities they read about (one was Tampa), the highs and lows in each city, the name of their own city, and the full name of the daughter who lives in each of the other four cities?

1. No two cities had the same high; no two cities had the same low; no two cities had the same difference between high and low.

2. These three cities had the highest highs (not necessarily in this order)—the city in which Mrs. Stone lives, the city in which Catherine lives, and San Francisco.

3. These three cities had the lowest lows (not necessarily in this order)—the city where Brenda lives, the city where Mrs. Valley lives, and Indianapolis.

4. The high was 10 degrees higher where Mrs. River lives than in Wichita; the low was 14 degrees higher where Mrs. River lives than in Wichita.

5. The city where Jack and Rebecca Hill live had a 5 degree lower low than the city where Felicia lives, which had a 5 degree lower low than Philadelphia had.

6. The high-low difference where Mrs. Forest lives was 1 degree less than in Philadelphia and 1 degree more than in the city where Mrs. River lives.

7. Of the five cities, the highest high was 28 degrees higher than the lowest high and 49 degrees higher than the lowest low.

8. The city with the highest high also had the highest low; the city with the lowest high also had the lowest low.

9. The Hills don't live in the city that had a high of 62° and a low of 42°.

10. None of the temperatures were higher than 88° or lower than 33°.

11. The low where Mrs. Forest lives was above 60°; Mrs. Forest is not Diane.

The solution is on page 161.

city	low	high	resident
___	___	___	___
___	___	___	___
___	___	___	___
___	___	___	___
___	___	___	___

CHALLENGER LOGIC PROBLEMS

66 RANKING THE PLAYERS

by Evelyn B. Rosenthal

A baseball team manager ranked Grant and six other players, #1 to #7, with #1 being best in both batting and fielding. The seven are the three basemen (1st, 2nd, and 3rd), the shortstop, and the three outfielders (left, center, and right). From this information and the clues, can you find the full name (one first name is Phil), position, and both rankings of each player?

1. Jim and the worst batter are both better fielders than Eames.

2. Clark, who is not Hal, is a better batter than the best fielder, who is not the #4 batter.

3. On both lists, the center fielder and 2nd baseman rank next to one another.

4. Right-fielder Archer is a better fielder than Ned, who is not the best batter.

5. In fielding, Brown is not the worst, but he is worse than Sam.

6. Outfielders are #1 on one list and #7 on the other; so are basemen.

7. Sam is a worse batter than Dunn, who is a worse batter than the worst fielder.

8. Fry, who is not the #4 fielder, does not play 1st base.

9. No two outfielders are next to each other on the fielding list.

10. Mel has the same rank on both lists.

11. Ned and Eames are, in some order, best and worst batter.

12. No two basemen are next to each other on the batting list.

13. Dunn and the #4 batter are, in some order, best and worst fielder.

14. Sam and Fry are ranked next to each other on both lists; so are Ken and Clark.

The solution is on page 162.

player	position	batting rank	fielding

113

67 THE HOUNDING AT BASKER HILLS

by Claudia Strong

Last week, two doctor friends, Elaine and Pat, went to see the hit thriller "The Hounding at Basker Hills." The movie began with six friends (one named Micah) moving into the accursed Basker Hills Castle. As the characters were besieged by increasing amounts of misfortune, Leatrice discovered a family history which told the story of the curse on the castle and outlined the way in which the curse might be defeated. Just as Leatrice was reading to the others about the six items hidden in six different rooms of the castle (including the dungeon), Pat's beeper went off, signaling that she had to return to the hospital at once. The next day, when Pat asked Elaine how the film ended, Elaine told her that the book Leatrice had found had gone on to explain that each of the characters had to find one of the magical items (which included a mystically engraved stone) and use them in that order to break the curse on the castle. Knowing that Pat loved Logic Problems, Elaine then handed her friend the following clues, from which Pat deduced the item each character found, the room in which each was found, and the order each was found and used to break the curse. Can you do the same?

1. Although the foyer contained weird trinkets, the magic potion was not found there.

2. Both the cloak and the password were found before Arden found his item.

3. When the sword was found, Duncan moved his search to another part of the castle and found the next item (which was not the cloak).

4. The final item used (which was not the staff) dispelled the evil spirits from the castle with a flash of lightning, which scorched the room in which it was found (not the dining hall), and knocked its user (who was not Arden) for a loop.

5. Rayna was sure that she would find the password in the tower, but after both that object and that room had been eliminated by two of the male residents, she did the rest of her searching in the foyer and the dining hall.

6. Gwendolyn found neither the cloak nor the item hidden in the presence chamber; she used her item third.

7. Leatrice found her item immediately before another of the women; neither of their objects were found in the kitchen.

8. The sword was not found in the tower.

9. The item which was used fourth was not the staff, nor was it found in the kitchen.

10. The staff was found after the item in the tower.

The solution is on page 162.

character	item	room	order

68 RIVERVIEW APARTMENTS

by Robert Nelson

The Riverview is a four-story apartment building with three apartments on each floor, as shown in the diagram. Of these twelve apartments, two are currently empty; the others are occupied by four married couples with no children, four married couples with one child each, a single woman, and a single man. One of the families' surname is Engel, one of the men is named William, and two of the children are named Amanda and Vince. From the following clues, can you determine the tenant(s) of each apartment?

1. One child lives on each floor, and there is at least one child in every line (A, B, or C) of apartments.

2. One of the two vacant apartments is immediately below the Wright apartment, and immediately above Roger's apartment; the other is immediately above the apartment Sam (a child) lives in.

3. Sally and her husband (who live two floors directly below a vacant apartment) are best friends with another couple on their floor, the Downings. Neither couple has children.

4. No adult's first or last initial corresponds with the letter of his or her apartment.

5. On one floor, the Carters live between June (who is not Bob's wife) and Mark; the Carters live two floors directly above Oscar (who is not Farrell).

6. Miss Hart sometimes babysits Faith, her next-door neighbor's child.

7. The following apartments are on the same floor: the Gordon apartment, Bob's, and Paula's; Mr. Alden's and Clara's; Zack's and the Farrell apartment; Laura's and Ed's.

8. Holly lives on a higher floor than both the Jacksons and Doris.

9. George has the same apartment letter as Tom, but he lives on a higher floor.

10. The two single people both live next to a vacant apartment.

11. Three of the children are Bob's child, the Brown child, and Nancy's son; Nancy and Ann live on the same floor, but Bob and Zack do not.

12. Zack and Mark are not next-door neighbors.

The solution is on page 163.

RIVERVIEW APTS.

69 LOGIC PROBLEM CONTEST

by David Champlin

Doll Crossword Puzzles recently ran a nationwide contest to discover new people with talent for creating Logic Problems. Contestants were invited to submit original Logic Problems, which were graded by the judges as easy, medium, or hard. Two winners were chosen from each level of difficulty, based on originality, humor, and how much fun the puzzle was to solve. When the winners were announced, the judges were pleased to announce that one puzzle had been received that was so difficult, original, and fun to solve that it had been awarded a special category, that of "Challenger," and awarded a special prize. Each of the seven winning entries was published in a different monthly issue of Doll Crosswords during one year, with the first winner appearing in the January issue. As it turned out, each of the winners, who included Ms. Nobel, lived in a different U.S. city, had a different occupation (one was a high-school track coach), and chose a different theme for his or her puzzle (one was about a classroom seating chart). From the information given below, determine each winner's full name, home city, occupation, the theme and difficulty level of his or her winning entry, and the order in which the puzzles were published.

Note: You may assume that the two winners for each level were of exactly equal difficulty.

1. No winning puzzles were published in any three consecutive monthly issues, nor were any published in May or June.

2. The winner from Milwaukee did not submit the puzzle about the track meet; the schoolteacher did not create the puzzle about the sweatshirts worn by a gym class.

3. Six of the seven winners were published in the following order from first to last: the puzzle about dolls (which was created by a woman); the puzzle written by the minister; the puzzle by Fern; the puzzle by the winner from Peoria; O'Reilly's puzzle; the medium puzzle published in November.

4. The puzzle submitted by the housewife from Denver was one of those about toys.

5. Ivan wrote one of the hard puzzles; the other hard puzzle was written by the winner from Little Rock (which was published one month before Ivan's).

6. Mr. Quincy, who is neither Ken nor Henry, and who is not from Milwaukee, wrote a puzzle easier than the puzzle published in July, which was not by Price.

7. Six of the seven winners are, in no particular order: the challenger creator; the pediatrician; the woman from Seattle; the one who created the puzzle about teddy bears; Mr. Minor; and the winner whose puzzle was published in April.

8. Greg does not live in Miami.

118

9. Neither Price (who does not live in Milwaukee) nor Rush lives in Peoria, Little Rock, or Seattle; their winning puzzles are about the track meet and the beauty contest (though not necessarily respectively).

10. Henry's puzzle was published in a winter issue; Ivan's was featured in a summer issue; two puzzles, one of which was Esther's, appeared in spring issues; Jane's puzzle and the puzzle about sweatshirts were published in the fall.

11. The librarian lives in neither Boston nor Peoria.

12. Esther (who is not Lewis) wrote a puzzle that was less difficult than the minister's; Henry's puzzle was more difficult than the pediatrician's, which was of equal difficulty to the puzzle about the shopping mall.

13. The sales manager lives in Miami.

The solution is on page 165.

The solution is on page 165.

first	last	occup.	city	topic	level	month

70 HANDCAR RALLY

by Randall L. Whipkey

The Summerset Handcar Rally is run annually over the old S. & O. C. Railroad train line to raise money for local charities. The route is divided into five legs going from one town to another, starting in Summerset and ending in Ocean City, with Mountainview one stop. This year, ten Summerset College freshmen raised over $1,000 in pledges by operating one of the handcars, with two different young men doing each rally leg. Given the information below, can you determine each pair of handcar pumpers, the towns in which they began and ended their rally leg, and the leg length?

1. The total distance of the rally was 120 miles. No two legs were the same length and all were double-digit distances.

2. The distance from Lakeside to Ocean City was 72 miles.

3. Herb and his partner pumped six more miles than Ed and his.

4. Brad and his partner pumped twice as far as the pair doing the fourth leg.

5. In consecutive order start-to-finish were these three legs: the one completed by Jeff and his partner, the one going from Lakeside to another town, and the leg ending in Riverfront.

6. Danny and his partner did six more miles than Ian and his partner, who isn't Jeff.

7. Ian and his partner did a later leg than Fran and his partner.

8. The second leg was half the distance of the leg completed by Chad and his partner.

9. These three legs were consecutive from first to last: the leg done by Greg and his partner, the 12-mile-long leg, and the one ending in Hilltop.

10. Herb and Adam, though roommates at the College, weren't partners.

11. Danny and Chad weren't partners.

The solution is on page 166.

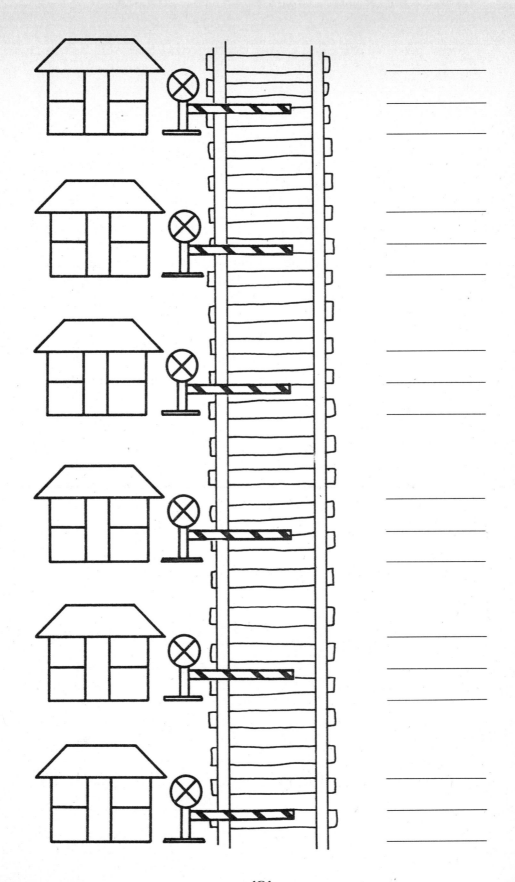

71 WEIRD PET RACES

by Claudia Strong

Six students who were avid fans of the late-night TV talk show hosted by Davey Mailcarrier, and who all had unusual pets, decided to hold their own "Weird Pet Races." Instead of the usual race, they found in an almanac the average speed of each of their pets. Each then timed his or her pet's speed, and rated their pets according to percentage by which each pet beat the almanac time. On the day of the race, the students were amazed to discover that all six pets beat the almanac's time; in fact, the sixth-place pet beat its average time by ten percent, while the winner of the race beat its average time by thirty or fewer percentage points. From the information below, can you determine the name of each student's pet, the type of animal it was, the percentage by which each beat its average time, and the order in which the pets finished?

1. Only two of the pets finished exactly one percentage point apart; these were the anteater (which was not Emma's pet) and Beth's pet.

2. Four of the animals were the aardvark, the armadillo, Baby, and the winner of the race.

3. Trisha's pet finished four points ahead of the chameleon, but behind pet Max.

4. The fifth-place pet finished ten points behind the third-place pet.

5. Three animals—Binky (which was not the armadillo), Chad's pet (which was not Star), and the second-place pet—placed in the 21- to 29-point range.

6. Buster finished directly behind (but seven points worse) than the platypus. Neither of these pets belonged to Larry or Beth.

7. Beth's pet finished in neither of the top two spots.

8. Three of the pets were Star (who was not Beth's), the iguana, and Corey's pet.

9. Larry's pet placed directly after Sparky.

The solution is on page 167.

student	pet	type	percent	order
				1
				2
				3
				4
				5
			10	6

72 SUMMERSET OUTLET MALL

by Randall L. Whipkey

The Summerset Outlet Mall consists of 12 shops, each offering a different line of goods to customers at discount prices; one shop sells pottery. As shown in the diagram on the next page, the mall is rectangular in shape, three shops wide and four shops deep. There are mall entrances in the two middle shops (shops 2 and 11) of the southern and northern ends of the building and each shop is connected to each shop adjoining it. Given this information and the clues that follow, can you deduce the name of each Summerset Outlet Mall shop, the kind of products it sells, and its location in the mall?

1. Of four shops, Ian's is as far from the small appliances outlet as possible, while Heidi's and the bath goods shop are also as far apart as possible.

2. The artwork outlet directly connects to Bill's and to the three shops that sell linens, baskets, and kitchenware.

3. Both the rug outlet and the Christmas shop can be entered from Leigh's.

4. From Jack's only three shops—Fran's, Gail's, and the glassware outlet—can be entered.

5. Carole's isn't the basket outlet.

6. Don's and the bath goods shop are not directly connected.

7. From Art's, only Don's and the candle outlet can be entered.

8. Ellen's, which is in #7, sells neither basketry nor kitchenware.

9. Ian's shop and the glassware shop aren't directly connected.

10. Gail's cannot be entered directly from Ian's.

11. Karen's sells discount light fixtures.

12. Heidi's sells neither rugs nor glassware.

13. Bill's and Jack's don't share a doorway.

14. Fran's doesn't sell artwork.

The solution is on page 168.

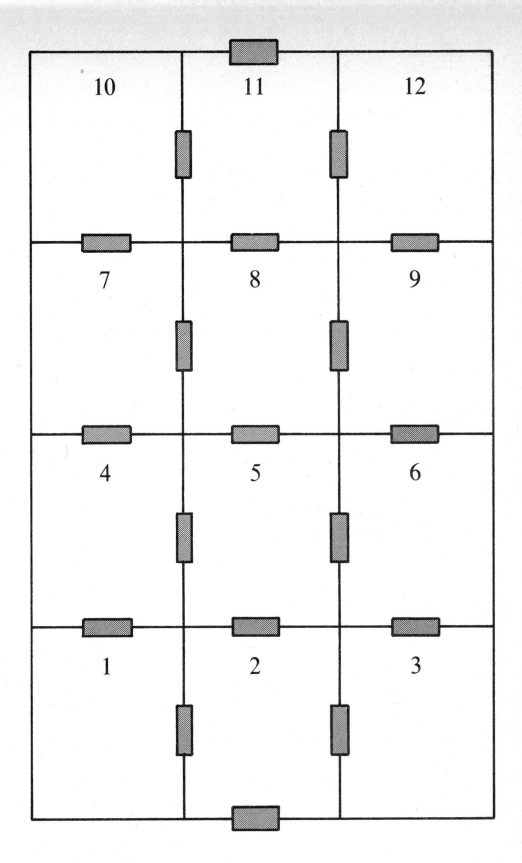

73 JURY DUTY

by Mary A. Powell

In Judge Moreland's court last Wednesday, a one-day traffic case was tried before a jury. The jury was selected from a list of 21 names, which included Michael and Ms. Thompson. The jury box was arranged with seats 1–6 (left to right) in the front row and 7–12 (left to right) in the back row, as shown in the diagram on the opposite page. As names were read off, in order, from the list, the person called for questioning took the first unoccupied jury seat. Those questioned were either accepted for jury duty, dismissed immediately, or dismissed after others had been questioned. From the following clues, can you find the full names (one last name is Varner) and professions of all 21 persons (one was a carpenter), the order of the list, and the jury seat of the final jurors? (Those who were dismissed left before the trial started.)

NOTE: Those who were dismissed after others had been questioned were not necessarily replaced immediately.

1. The first three on the list were accepted for the final jury, as was the last person (#21) on the list, who was juror #12.

2. When the trial started, Ruth sat directly behind Ms. Olson, the electrician sat directly behind Clarence, and Mr. Friedman sat directly behind Gloria.

3. Of these four final jurors, Richard and the hotel clerk sat as far apart as possible, as did Iris and Johnson.

4. The florist was accepted for the final jury; the veterinarian was dismissed immediately; Dorothy and the bus driver were each dismissed after at least five others had been questioned.

5. These three were on the list consecutively: Patricia, the teacher, and Moore. One of them was dismissed.

6. The first man on the list was the auto mechanic; the last man on the list was Anthony. The first woman on the list was Alice; the last woman on the list was Ms. Channing.

7. The first three to be dismissed were Mr. Braden, Jennifer, and the jeweler, in that order.

8. The last three to be accepted for the final jury were the photographer, Mr. Dreyer, and the nurse, in that order.

9. Norman's name was on the list immediately before Gloria's and immediately after the chiropractor's.

10. Catherine was #8 on the list; Mr. King was #16.

11. When the jury filed out in numerical order for lunch, Lovell was directly behind the antiques dealer and directly in front of the musician, who is a man.

12. Three final jurors were, in consecutive order, Ms. Young, Robert, and the waiter.

13. The accountant's jury seat number was higher than Mr. Rothman's.

14. Four women's names were listed immediately before Larry's, and three other women's names were listed immediately after Larry's.

15. After the trial began, two men were in the front row of the jury box; the other three were in the back row.

16. Ms. Waters was listed between Ms. North and Mr. Simmons; after the trial began, she was seated between Catherine and Mr. Rothman. (Five people are mentioned here.)

17. The housewife was called for questioning immediately after Ms. Fox and immediately before Susan.

18. After the trial started, Margaret sat in the front row between the realtor and Mr. Glover.

19. The woman dietitian (who is not Ms. Olson), was listed before the computer programmer, who was listed before Ms. Potter (who is not the housewife).

20. After the trial started, Jerome sat beside the man who sat behind Mr. Rothman, who sat beside Henry (Jerome did not sit behind Henry).

21. William's name was on the list immediately before that of Wilkins, who was listed before the office manager.

22. Frank, who is not Mr. King, sat in the jury box while two others were questioned; Mr. Quigley was dismissed after Frank was dismissed.

The solution is on page 169.

The solution is on page 169.

7	8	9	10	11	12
1	2	3	4	5	6

The fill-in chart for keeping track of the list order is on page 128. You might find it more comfortable to copy the chart onto a separate piece of paper.

The fill-in chart for keeping track of the list order is on page 128.

list order	first	last	occup.	accept/ dismiss	seat #

74   BASKETBALL FIVE

by Evelyn B. Rosenthal

In basketball, a free throw scores 1 point, some field goals score 2 points and some score 3. In a recent high-school game, Dan and four others scored a total of 71 points. From this information and the clues, can you find the full name of each boy, his position, how many 1-, 2-, and 3-point shots he made, and his total number of points?

1. Smith had twice as many 2-point shots as Bob, and Chuck had twice as many 2-point shots as Rowe.

2. Only Al and Phelps, who is not the center, had total scores within a point of the average of all five boys.

3. Wolfe scored as many 2-point field goals as Chuck scored free throws (1-point goals), and Tate scored as many 2-point field goals as Ed scored free throws.

4. The two guards had the lowest total scores; one of the forwards had the highest.

5. The player who had the largest number of 3-point scores (two more than the player who had the fewest), is Bob (who is not Rowe).

6. Except for a forward, who had the same number of 1- and 2-point shots, no one had the same number of any two of the three kinds.

7. The number of 2-point shots was four more than the number of free throws scored.

8. Each boy scored at least one 2-point shot; no two scored the same number.

9. Each boy scored at least one free throw.

10. Each total score was, at most, two points away from the one closest to it.

The solution is on page 171.

first name	last name	position	1 pt.	2 pt.	3 pt.	total

FARMERS' MARKET

by Mary A. Powell

When Woodland's first farmers' market opened last Tuesday, the first five customers loved the fresh produce and were glad to have an outdoor market so close to home. Between them, they bought everything the twelve sellers had to offer. From the following clues, can you find the full names of the first five customers, the types of produce each bought, and the table number where each item was sold?

1. Each vendor offered only two types of produce, but none sold the same two items.

2. Each customer bought from five different sellers; no one bought more than seven types of produce, and no one bought the same type of produce from different sellers.

3. Sellers #3, #6, and #8 each sold apples; Karen, Loretta, and Ms. Quincy were the three who bought apples.

4. Sellers #3 and #7 each sold walnuts; Marcia and Ms. Smith were the two who bought walnuts.

5. Noreen and Ms. Thompson both bought corn from seller #1; neither was the woman who bought a basket of raspberries.

6. Janet and Noreen bought more items than Ms. Quincy and Ms. Purcell, but fewer than Ms. Rush, who bought from three corner vendors.

7. The woman who bought yellow onions did not buy anything at tables #9 or #11.

8. The woman who bought lettuce did not shop at tables #2 or #12. Lettuce was not sold at a corner table.

9. The woman who bought a watermelon did not shop at tables #7 or #11; she and one of the women who bought a pumpkin also bought both nuts and berries.

10. Ms. Purcell, who is not Karen, did not buy anything at table #3.

11. The squash sellers were both adjacent to the same corner table; neither sold cantaloupes or tomatoes.

12. Sellers #5, #6, and #10 had only one customer each.

13. The seller who sold green beans and carrots had a higher table number than the one who sold yellow onions and carrots.

14. The two almond vendors were at corner tables.

15. The seller who sold pumpkins had table #11 on one side and the seller who sold green onions on the other.

16. Three customers bought strawberries; the others were the two who bought from seller #9.

17. One tomato seller was adjacent to a cantaloupe seller; the other tomato seller was adjacent to a squash seller. Tomatoes were not sold at tables #5, #6, or any corner table.

18. Janet and Ms. Rush bought no almonds. Noreen didn't buy strawberries.

19. Neither of the two cantaloupe sellers was at a corner table and neither sold raspberries, lettuce, or tomatoes.

20. Karen, who didn't buy squash, was not one of the two customers at table #11.

21. Three women bought tomatoes; one of these three also bought squash.

22. Berries were not sold at table #7.

The solution is on page 173.

1	2	3	4
12			5
11			6
10	9	8	7

first				
last				
items bought				

SOLUTIONS

1. FOUR A's ANTIQUES

The last to arrive was not Amy (clue 1), Anna (clue 2), or Ada (clue 3); so she was Ava, whose specialty is antique books (clue 4). The first to arrive was not Ms. Trent (clue 1), Ms Dent (clue 2), or Ms. Brent (clue 3), so she was Ms. Kent. By clue 1, the jewelry expert arrived third, Ms. Trent arrived second, and Amy Kent arrived first. Also by clue 1, Ms. Trent is not the furniture expert, so Amy Kent is; by elimination, Ms. Trent's specialty is antique clothing. By clue 2, then, Anna arrived third and Ms. Dent arrived last. By elimination, Ada arrived second; also by elimination, Ms. Brent arrived third. In sum, in order of arrival:

> Amy Kent, furniture
> Ada Trent, clothing
> Anna Brent, jewelry
> Ava Dent, books

2. IT'S DONE WITH MIRRORS

Downing is not Nicole (clue 2) or Lloyd (clue 4). Downing's mirror is not round (clue 2) or oval (clue 7), so it is either hexagonal or rectangular. Myra's mirror is not hexagonal (clue 3) or rectangular (clue 6), so Myra is not Downing; Oscar is. Oscar's mirror was not $350 (clue 1), $175 (clue 1), or $300 (clue 2), so it was $225. The oval mirror cost $175 (clue 1). The round mirror did not cost $225 (clue 2) or $350 (clue 2), so it cost $300. Nicole's mirror cost $350 (clue 2). The rectangular mirror did not cost $350 (clue 6), so it cost $225 and the hexagonal mirror cost $350. Evans's mirror did not cost $175 (clue 1) or $350 (clue 5), so it cost $300. Fairfax's mirror did not cost $175 (clue 6), so it cost $350 and, by elimination, Glass's mirror cost $175. Myra is not Glass (clue 3), so she is Evans and, by elimination, Lloyd is Glass. In summary, in order of price:

> Lloyd Glass, oval, $175
> Oscar Downing, rectangular, $225
> Myra Evans, round, $300
> Nicole Fairfax, hexagonal, $350

3. ALL THE KING'S HORSES

The six horses were as follows: Prince, the sorrel gelding (clue 1); Duke, ridden by Sue (clue 1); the Appaloosa, ridden by Eve (clue 2); Thor, the black stallion (clue 3); the pinto ridden by Nan (clue 3); and Pixie ridden by Pat (clue 4). Since the horses were from 3 to 6 years old (introduction), Thor must be 6 years old and Nan's pinto 3 years old (clue 3). Pat's Pixie and Tomboy are 4 years old (clue 4); by elimination, Prince (the sorrel) and Duke (who was ridden by Sue) are 5 years old (clue 5). Tomboy must be the Appaloosa ridden by Eve. Thor, the black stallion mentioned in clue 3, was not ridden by Bea (clue 6), so he was ridden by Kay and Bea rode Prince. By elimination, Nan's pinto was Angel. Lastly, according to clue 7, 5-year-old Duke is the chestnut and Pixie is white. In sum:

> Bea, Prince, sorrel, 5
> Eve, Tomboy, Appaloosa, 4
> Kay, Thor, black, 6
> Nan, Angel, pinto, 3
> Pat, Pixie, white, 4
> Sue, Duke, chestnut, 5

4. MEDITERRANEAN WINDS

Tom's name is Davis (clue 4). Adam's is not Price (clue 1), Smith (clue 3), or Newton (clues 1 and 5), therefore it is Jones. Hiram's surname is not Newton (clue 5), or Price (clues 1 and 6), therefore it is Smith. Abel's surname cannot be Price (clues 1 and 6) therefore it is Newton and by elimination, Dick's surname is Price. By clue 5, Abel Newton's ship is not a brigantine and by clue 6 it is not a brig, so his ship is the bark. By clue 2, it went to Split for repair of damage caused by a bora. By clue 7, Adam Jones's ship was repaired at Barcelona. By clue 3, Hiram Smith's ship was repaired at Marseilles. By clue 4, Tom Davis' ship was damaged by a sirocco. By clue 8, it couldn't be the ship repaired at Valetta so, by elimination, it was repaired at Tripoli and Dick Price's ship at Valetta for damage caused by a gregale. By clue 1, Adam Jones's ship was not the one damaged by a mistral, so by elimination, it was damaged by a lebeche, and Hiram's ship, then, by a mistral. By clue 1 we know Dick Price served on one of the two brigs, so it was the one that was damaged by a gregale and repaired at Valetta. The other brig then, was Tom Davis's ship which was damaged by a sirocco and repaired at Tripoli. By clue 1, Adam Jones's ship was a brigantine and was the one damaged by a lebeche and was repaired at Barcelona. Hiram, by elimination, served in the brigantine damaged by the mistral and repaired at Marseilles. In sum:

> Tom Davis, brig, sirocco, Tripoli
> Adam Jones, brigantine, lebeche, Barcelona
> Abel Newton, bark, bora, Split
> Dick Price, brig, gregale, Valetta
> Hiram Smith, brigantine, mistral, Marseilles

5. COUNTY FAIR COWS

The first-place owner was not Sal (clue 1), Al, or Hal (clue 2); he was Cal. The fourth-place owner was not Sal (clue 1) or Al (clue 2); he was Hal. By clue 2, therefore, Al's cow placed third; by elimination, Sal's placed second. Cal is Mr. McGee (clue 1); Sal is Mr. Magee (clue 2). By clue 4, Hal is not Mr. McFee, so Al is; by elimination, Hal is Mr. McKay. By clue 3, Blue Streak placed fourth. By clue 1, then, Thunderbolt placed third. Also by clue 1, Lightning placed first; by elimination, Rainbow placed second. In sum:

> First prize, Cal McGee, Lightning
> Second prize, Sal Magee, Rainbow
> Third prize, Al McFee, Thunderbolt
> Fourth prize, Hal McKay, Blue Streak

6. HAWAIIAN HOLIDAY

The woman who is going snorkeling is not Lauren (clue 2), so she is Jennifer. The sunbather is not Lauren (clue 1) or Michael (clue 6); he is Brett. Michael is not the sightseer (clues 5 and 6), so Lauren is and, by elimination, Michael is the shopper. The person with Multitour is not Brett (clue 6), Lauren (clues 5 and 6), or Michael (clue 6); she is Jennifer. The men will be on Kauai (clue 2) and Maui (clue 4). Jennifer will not be on the Big Island (clue 3); she will be on Oahu and Lauren will be on the Big Island. The Worldwide person is not Brett (clue 1) or Lauren (clue 7); he is Michael. Lauren is not on the Tampos tour (clue 5); she is with Carvas and, by elimination, Brett is with Tampos. Brett will not be on Maui (clue 4), so he will be on Kauai and, by elimination, Michael will be on Maui. In summary:

> Brett, sunbathing, Kauai, Tampos
> Jennifer, snorkeling, Oahu, Multitour
> Lauren, sightseeing, Big Island, Carvas
> Michael, shopping, Maui, Worldwide

7. BIRTHDAY GREETINGS

Harry is not Mr. Wilcox (clue 1); Michael is. Callers included a niece from California, an uncle from Alaska (clue 2), and a former roommate who called from Hawaii (clue 4). Of the three relatives, one is from Ohio (clue 5), so must be the grandmother (clue 2). By elimination, the teacher is from Florida. Alice, then, is the niece from California and

Peterson is the uncle from Alaska (clues 2, 5), who must be Harry. Michael, then, is the former college roommate from Hawaii. The Florida call came before Michael's (clue 1); Leona's came after Michael's (clue 3), so Leona is from Ohio and Jane is the teacher from Florida. McMillan's call came before Michael's (clue 4), so Leona is not McMillan or Grayson (clue 3); Leona is Denver. Jane is not McMillan (clue 4), so Jane is Grayson and Alice is McMillan. In summary:

> Leona Denver, grandmother, Ohio
> Jane Grayson, teacher, Florida
> Alice McMillan, niece, California
> Harry Peterson, uncle, Alaska
> Michael Wilcox, roommate, Hawaii

8. MUSIC LESSONS

The first student is not Frohman (clue 1), Freedman (clue 2), or Friedman (clue 3), so Freeman is. The second student is neither Frohman (clue 1) nor Freedman (clue 2), so Friedman is. The first student is not John (clue 1), Elaine (clue 2), or Nancy (clue 3); he is Ted. The last student is neither John (clue 1) nor Elaine (clue 2); she is Nancy. By clue 2, Freedman is a man, but he is not the first student; since he is not Ted, he must be John. Nancy is not Friedman (clue 3), so Elaine is; by elimination, Nancy is Frohman. Also by elimination, John Freedman must be the third student. By clue 2, Ted studies piano. By clue 1, Elaine studies guitar. Finally, by clue 4, John does not study viola, so Nancy does; by elimination, John studies violin. In sum, from first to last:

> Ted Freeman, piano
> Elaine Friedman, guitar
> John Freedman, violin
> Nancy Frohman, viola

9. THE KNITTERS

The three women who took lessons were Anna Franklin, Jones, and the one who made the sweater (clue 2); since Clara took lessons but did not make the sweater (clue 5), she must be Jones. Ms. Harris made the red item (clue 7); she and Ms. Iverson were the two who didn't take classes (clue 2). By elimination, Ms. Graves made the sweater. Donna made the blue blanket (clue 4). She is not Harris (clue 7) or Graves, who made the sweater; she is Iverson. The woman who made the hat took knitting lessons and is not Jones (clue 6), so she is Anna Franklin. Since Edith did not make the red item (clue 2), she is Graves. By elimination, Betty is Harris. The mittens were green (clue 1). The only woman who could have made them is Clara Jones. By elimination, Harris made the scarf. Clara Jones previously knew the one who used white yarn (clue 3), who is not the one who made the hat (clue 6); the sweater must be white and the hat yellow. In sum:

> Anna Franklin, yellow hat
> Edith Graves, white sweater
> Betty Harris, red scarf
> Donna Iverson, blue blanket
> Clara Jones, green mittens

10. VALENTINE WEDDINGS

Mr. McKinney is not marrying Patty, Ruth (clue 3), or Tammy (clue 6); he is marrying Susan. Susan's last name is not Gill (clue 1) or Neal (clue 6); nor can it be Kinney (intro), so she is Susan Dowell. Tammy is not Ms. Gill (clue 1) or Ms. Neal (clue 6); she is Ms. Kinney. Jerry's fiancee is not Susan Dowell, Tammy Kinney, or Ms. Neal (clue 6); she is Ms. Gill. Jerry's last name is not McDowell (clue 4), nor can it be McGill; he is Jerry McNeal. Mr. McGill is not Larry or Mike (clue 5); he is Keith. Ms. Neal is not engaged to Mr. McGill (clue 2); nor can she be engaged to Mr. McNeal—she is engaged to Mr. McDowell. By elimination, Ms. Kinney is engaged to Mr. McGill. McDowell is not engaged to Patty (clue 4), so he is engaged to Ruth. His name is not Larry (clue 7); he is Mike. By elimination, Patty is Ms. Gill, and Larry is Mr. McKinney. In summary the couples are:

Susan Dowell and Larry McKinney
Patty Gill and Jerry McNeal
Tammy Kinney and Keith McGill
Ruth Neal and Mike McDowell

11. GREETING CARD PARTNERS

The man with the most experience is not Mr. Malden (clue 1), Moore (clue 4), or Mercer (clue 6); he is Mr. Mason. He is not Eugene (clue 1), Barry (clue 2), or Arthur (clue 3), so he is Harold. The man with the least experience is neither Eugene (clue 1) nor Barry (clue 7), so he is Arthur, with four years' experience (clue 3); by clue 7, therefore, Barry has the second-highest years of experience and Eugene has the third-highest. Arthur must be Mr. Malden (clue 1). Also by clue 1, Eugene has eight years' experience, while "Chip" has sixteen. By the introduction, then, Harold Mason has twenty years' experience; by clue 4, he must be "Buzz" and "Chip" with sixteen years is Barry. By clue 1, Barry is not Mr. Mercer, so Eugene is; by elimination, Barry is Mr. Moore. By clue 2, "Flip" must be Arthur Malden; by elimination, Eugene Mercer is "Tank." By clue 6, the printer is Harold Mason. By clue 5, then, the designer is Barry Moore. Finally, the packager must be Eugene Mercer (clue 8); by elimination, Arthur Malden is the shipper. In sum:

Harold "Buzz" Mason, printer, 20 years
Barry "Chip" Moore, designer, 16 years
Eugene "Tank" Mercer, packager, 8 years
Arthur "Flip" Malden, shipper, 4 years

12. MILE RELAY TEAM

Since all four team members are listed in clue 1, one must be the second-leg runner. He is not Gary (clue 4) or the boy who ran in 52 seconds (clue 6). Utley then ran the second leg. Ivan did not run the first, second, or third leg (clue 5); he was in fourth position. Gary did not run the second or third leg (clue 1); he ran the first leg. The boy who ran in 52 seconds must be Ivan (clue 1). Gary was neither the fastest nor slowest runner (clue 2). His time must have been 49 seconds. Henry was not the fastest runner (clue 2); his time was 53 seconds. By elimination, Jeff ran in 48 seconds. The second runner was not the fastest (clue 4); his time was 53 seconds, and he is Henry. By elimination, the third runner was the fastest, Jeff is not Tyler or Robbins (clue 3); he is Shelton. Since Tyler ran before Robbins (clue 3), Tyler must be Gary, while Ivan is Robbins. In summary:

1st leg: Gary Tyler, 49 seconds
2nd leg: Henry Utley, 53 seconds
3rd leg: Jeff Shelton, 48 seconds
4th leg: Ivan Robbins, 52 seconds.

13. THE PINEVALE BUGLE

By clue 8, the letter favoring commercial zoning was published Monday. From the introduction and clue 2, Rose Dorgan's letter, which favored residential zoning, was published on Wednesday. By clue 3, Jackson wanted the land to be used for a park, so his letter could not be the one published Monday; as his letter was published one day earlier than King's it could not have been published on Tuesday (as this would cause King's letter to be published Wednesday). By clue 1, King's letter couldn't have been published Saturday; by elimination, Jackson's letter was published Thursday and King's Friday. By clue 5, Fred's letter was published Thursday so his surname is Jackson. By clue 1, the letter which advocated industrial zoning could not have been King's and it couldn't have been the one published

Saturday; by elimination, it was published Tuesday. By clue 6, Floyd's letter, which advocated using the land for schools, could not be the one published Saturday, so by elimination, it was published Friday. Also by elimination, the letter recommending agricultural zoning was published Saturday, which, by clue 4, was Dolores' letter. By clue 6, Floyd's letter was published before Lopez's, therefore Lopez's letter was published on Saturday. By clue 1, Sarah cannot be the one favoring industrial zoning, so she favors commercial zoning and, by elimination, Dan favors industrial zoning. By clue 7, Dan's surname is not Brown, therefore it is Curtin and Sarah's is Brown. In sum:

> Monday: Sarah Brown; commercial
> Tuesday: Dan Curtin; industrial
> Wednesday: Rose Dorgan; residential
> Thursday: Fred Jackson; park
> Friday: Floyd King; schools
> Saturday: Dolores Lopez; agriculture

14. EASTER EGG HUNT

Lawson is not Laura (clue 1), Chris, Greg (clue 3), or Stephen (clue 4); Lawson is Mandy. Grant is not Greg (clue 1), Chris (clue 3), or Stephen (clue 4); she is Laura. Stephen is not Smith or Star (clue 1); he is Farmer. Chris is older than Greg (clue 3). Therefore, since the Star child is the oldest (clue 5), Greg is not Star; Greg is Smith, and by elimination, Chris is Star. The six-year-old is not Stephen, Laura Grant (clue 4), or Chris, who is the oldest (clue 5); the seven-year-old found fifteen eggs (clue 6); Greg Smith is not the seven-year-old (clue 2). If Greg is not seven, then Mandy Lawson, one year his junior (clue 3), can't be the six-year-old. The six-year-old is Greg Smith. Mandy Lawson is five years old (clue 3). Chris is ten years old (clue 3). Greg found twelve eggs (clue 2). Stephen found fifteen eggs and Laura Grant found eighteen (clue 4). Mandy found thirteen eggs (clue 7). Stephen is the seven-year-old (clue 6). By elimination, Laura is the eight-year-old. Chris, then, found seventeen eggs (clue 5). In summary:

> Chris Star, 10 years old, 17 eggs
> Laura Grant, 8 years old, 18 eggs
> Stephen Farmer, 7 years old, 15 eggs
> Greg Smith, 6 years old, 12 eggs
> Mandy Lawson, 5 years old, 13 eggs

15. RETRAINING

By clue 1, the five workers are Tom, the one who used to be a welder, the arcade manager, Ms. Cortez, and the fitness instructor. Ralph used to be a foreman (clue 2), so he is either the arcade manager or the fitness instructor; the same is true of Marie, the former secretary, who is not Ms. Cortez (clue 3). Since there are only two men among the five, Mr. Hampton, the mail carrier (clue 4), must be Tom. Ralph is then Mr. Bertram, who is not the fitness instructor (clue 1), so he is the arcade manager, and Marie is the fitness instructor. The programmer, who used to repair TVs (clue 5), must be Ms. Cortez, and her first name is not Erica (also clue 5), so she is Sylvia, while Erica is the former welder. Monroe is not Marie (clue 3), so she is Erica. By elimination, Tom used to be a telephone operator, Erica is now a mechanic, and Marie's last name is Swanson. In sum, with former job first and present job second:

> Ralph Bertram: foreman, arcade manager
> Sylvia Cortez: TV repairer, programmer
> Tom Hampton: phone operator, mail carrier
> Erica Monroe: welder, mechanic
> Marie Swanson: secretary, fitness instructor

There are two boys and two girls mentioned. By clue 5, Steve must be the boy who likes red lights, and he is three years younger than Sam. By clue 3, then, Sam must be the boy who likes green lights and he's two years younger than Sally, while Susan must be the girl who likes a Santa's elf ornament, and she is half Sally's age. By clue 1, one child, not the youngest, is seven years old. Since Sally is twice someone's age, her age is an even number, so, in clue 3, neither Sally nor Sam is the seven-year-old. If Susan were the seven-year-old, then Sally would be fourteen, Sam twelve, and Steve nine, and the seven-year-old would be the youngest, contradicting clue 1. Therefore, Steve is the seven-year-old, Sam is ten, Sally is twelve, and Susan is six. By clue 4, Sally must be the one who likes blue lights and, by elimination, Susan likes gold lights. By clue 2, Sam must be the one who likes the cardinal ornament. By clue 4, Steve, then, likes the snowman ornament and, by elimination, Sally likes the star. In sum:

> Sally, 12: blue, star
> Sam, 10: green, cardinal
> Steve, 7: red, snowman
> Susan, 6: gold, Santa's elf

17. WOODLAND WATERING SPOT

The group of two animals came to the spring first, not by the path or the old oak tree (clue 4); nor did they come through the tall grass (clue 2), so they came over the rock. By clue 3, the ravens and partridges together must have numbered nine (one group of four and one of five) in order to be three times the number in one group. By clue 5, the raccoons didn't approach over the rock, so the group of two were the pumas, and the raccoons numbered three. The third group to arrive at the spring were the partridges (clue 6). By clue 5, the three raccoons weren't last, so they were second and the ravens were last. By clue 2, there were four ravens, who came through the grass; there were then five partridges. The partridges didn't come by way of the oak tree (clue 1), so they used the path. The raccoons, then, came by way of the tree. In sum, in order of arrival:

> 2 pumas, over the rock
> 3 raccoons, oak tree
> 5 partridges, by path
> 4 ravens, through grass

18. MOTHER'S DAY GIFTS

By clue 4, the Polar child is ten years old, Rusty is nine, and the girl who bought the candy bar eight. A girl bought the candy bar (clue 4), but not Beth (clue 6), so Dotty did. By clue 1, Carl, who is one year younger than Dotty, is seven, while the Brook child, who isn't Rusty, must be a second nine-year-old. By clue 2, the Smith child must be either Rusty or Dotty, and in either case, Carl bought the bubble bath. The Heart child who bought the puzzle magazine (clue 3) can only be Rusty, so the Smith child is Dotty; Carl's last name, by elimination, is Jones. The Brook child is then Beth (clue 5), while the Polar child is Tom. Beth didn't buy the chocolate-covered cherries (clue 6), so Tom did, and Beth bought the nail polish. In sum:

> Tom Polar, 10: cherries
> Beth Brook, 9: nail polish
> Rusty Heart, 9: puzzle magazine
> Dotty Smith, 8: candy bar
> Carl Jones, 7: bubble bath

19. THE COFFEE SHOPPERS

The most expensive coffee was from Celebes (clue 4). The woman who bought it (clue 5) is not Sue (clue 4) or Beth (clue 2); she is Meg. The least expensive Excel coffee was bought by a man, but not Ian (clue 8), or Leo (clue 2), so Joe bought Excel. Leo's coffee variety was strong (clue 3), so, by clue 7, Joe selected the Colombian blend and the Estate variety was Ian's choice. Leo chose the Peaberry variety (clue 5). By clue 6, Mr. Frey chose Java coffee and is not Leo; he is Ian. Leo's strong variety isn't from Sumatra (clue 3) or China (clue 6); it is from Kenya. Mr. Dunn didn't buy Excel (clue 8), so he is Leo. The inexpensive Yunnan variety isn't from Sumatra (clue 9) or Celebes (clue 4), so it is from China. Since Beth bought the second most expensive coffee (clue 2) Sue bought Yunnan. The Kalossi variety isn't from Sumatra (clue 3), so it is from Celebes, while the Lintong variety must be from Sumatra, and it was Beth's choice. By clue 7, Ms. Ward, who didn't buy the strong varieties mentioned in clue 3, is Sue. Ms. King isn't Meg (clue 1), so she is Beth. Smith didn't buy coffee from Celebes (clue 4), so Smith isn't Meg, and must be Joe. By elimination, Meg is Long. In summary:

> Leo Dunn: Kenya, Peaberry
> Ian Frey: Java, Estate
> Beth King: Sumatra, Lintong
> Meg Long: Celebes, Kalossi
> Joe Smith: Colombia, Excel
> Sue Ward: China, Yunnan

20. AIRPORT ACQUAINTANCES

A woman sat in Seat #3 (clue 4). She is not Mr. Owens (introduction), Mr. Quinlan (clue 1), Mills (clue 2), or Nolan (clue 3); she is Ms. Price. She was on a business trip (clue 6), but not to Alaska (clue 5), so she was going to New York (clue 1). She is not Karen (clue 1), so she is Gertrude, the only other woman at the table. Mills was not going to Alaska, Colorado, or Hawaii (clue 7), so Mills was going to Pennsylvania. Karen was on a vacation trip (clue 8), but not to Hawaii (clue 1), so she was going to Colorado. She cannot be Mr. Owens (introduction) or Mr. Quinlan (clue 1), so she is Ms. Nolan. Mr. Quinlan was not going to Hawaii (clue 1), so he was going to Alaska; by elimination, Mr. Owens was going to Hawaii. By clue 1, Mills must be Ira. By clue 5, Karen Nolan sat in either Seat #1 or Seat #5. The same holds true for Jim (clue 3). Jim did not sit in Seat #1 (clue 9), so Karen Nolan did, and Jim sat in Seat #5 (clue 3). Hank sat in Seat #4 (clue 9); by elimination, Ira sat in Seat #2. Jim is Mr. Quinlan (clue 5); by elimination, Hank is Mr. Owens. In sum:

> Seat #1, Karen Nolan, going on vacation to Colorado
> Seat #2, Ira Mills, headed to college in Pennsylvania
> Seat #3, Gertrude Price, business trip to New York
> Seat #4, Hank Owens, going on vacation to Hawaii
> Seat #5, Jim Quinlan, business trip to Alaska

21. SHOWTIME

Ursula attended the Bijou (clue 1). The show at the Uptown was not attended by Rita or Teresa (clue 1), Sally (clue 2), or Wanda (clue 3); it was attended by Vanessa, whose surname is King (clue 1). Wanda went to the show at Cinema 2, and the Cinema 1 show was attended by the Sampsons (clue 3). Ed attended the show at Cinema 4 (clue 4). It must have been with Rita while Andrew went to Cinema 3 (clue 3) at 9 p.m. (clue 2). Andrew didn't, then, go with Sally who went to the mall for the 7 p.m. show (clue 2); she must be Mrs. Sampson in Cinema 1 and, by elimination, Teresa attended Cinema 3 with Andrew. Neither Rita nor Teresa is Hughes (clue 1); thus, the Hughes couple attended Cinema 2, and Doug went to Cinema 1. Comedies were shown at the Uptown and Bijou (clue 1), so by clue 2, Chuck was at Cinema 2 at 7 p.m. The two shows not at the mall both began at 7 p.m. (clue 5). The Jones couple who made a 9 p.m. show (clue 2) must be Rita and Ed in Cinema 4. By clue 2, Green attended the 7 p.m. show at the Bijou and Bob went to the Uptown (clue 2).

By elimination, Frank went to the Bijou with Ursula and the Taylors attended Cinema 3. In sum:

> Doug and Sally Sampson, Cinema 1, 7 p.m.
> Chuck and Wanda Hughes Cinema 2, 7 p.m.
> Andrew and Teresa Taylor, Cinema 3, 9 p.m.
> Ed and Rita Jones, Cinema 4, 9 p.m.
> Frank and Ursula Green, Bijou, 7 p.m.
> Bob and Vanessa King, Uptown, 7 p.m.

22. THE ENLISTEES

By clue 1, the first two graduates were John and the young man who went into the Marines; the second set of two graduates joined the Army; the radio repairer is still another graduate. By clue 4, Pat joined the Air Force and is not the radio repairer, and so is yet another graduate. John and the radio repairer must have joined the Navy (clue 3). By clue 2, Reginald, an auto mechanic, was not in the Army, so he was the Marine. Willy the carpenter (clue 4) and Dana (clue 3) joined the Army, and, by elimination, Kim was the Navy radio repairer. John was not the dental technician (clue 6) or the stenographer (clue 5); he was the electrician. Pat was not the dental technician (clue 4), so Dana was, and Pat was the stenographer. The last to enlist was either Kim or Pat (clue 1): it was not Kim (clue 7) and so it was Pat, whose last name is then Martin (introduction). By clues 4 and 5, Willy and John are her brothers; the Fremonts, therefore, are Dana, Kim, and Reginald. In sum:

> Dana Fremont: Army, dental technician
> Kim Fremont: Navy, radio repairer
> Reginald Fremont: Marines, auto mechanic
> John Martin: Navy, electrician
> Pat Martin: Air Force, stenographer
> Willy Martin: Army, carpenter

23. MIXED-UP COFFEE MUGS

By clue 3, Julie picked up Donna's mug, so Donna's was not the wise-saying mug (clue 2). Nor, since it was on Julie's hook on the top row, was it the initial mug (clue 1) or the striped mug (clue 5). Since the striped and initial mugs belong on the top row, they belong, in one order or another, to Julie and George. The wise-saying mug, which belongs to a woman (clue 2), then, must belong to Phyllis. The initial mug does not hang directly above it (clue 1), so the initial mug could now only belong to George, while the plain mug must belong to Chad (also clue 1). By elimination, Julie's mug was the striped one, which by clue 5, was mistakenly placed on Phyllis' hook. Using clue 4, the only remaining corner mug which could move from corner to corner is Maxwell's; his college-logo mug must have ended up on George's hook; by elimination, Donna owns the rainbow mug. By clue 6, the only two men who could exchange mugs are George and Maxwell. Chad was then the man who picked up Phyllis' wise-saying mug (clue 2) and by elimination, Chad's plain mug was mistakenly placed on Donna's hook. In sum, we have:

> Donna, owned rainbow mug, picked up Chad's
> Julie, owned striped mug, picked up Donna's
> George, owned initial mug, picked up Maxwell's
> Maxwell, owned college-logo mug, picked up George's
> Phyllis, owned wise-saying mug, picked up Julie's
> Chad, owned plain mug, picked up Phyllis'

24. RELICS OF THE PAST

By clue 2, there were three items donated the second week; two household items were gifts from two of the three women—one of which was the fourth donation. The three consecutive

140

donations mentioned in clue 4 of a wagon, and Mr. Jacobs' gift, and the trunk, therefore, could not all have been the last three gifts. Since the fourth gift was a woman's household item, Mr. Jacobs' donation wasn't fourth. The wagon wasn't first (clue 4), so the wagon was second, Jacobs' gift third and the trunk fourth. The three donations mentioned in clue 5 are also consecutive, and none of them can be third—Mr. Jacobs holds that place—so, they must be fourth, fifth, and sixth donors, i.e., the woman from Senton gave the trunk fourth, Mr. Reyes is fifth, and Ms. Alcott is sixth. Ms. Miller isn't from Senton (clue 5), so, by elimination, Ms. Beebe is. Mr. Reyes, who is fifth, gave the stove (clue 3); he is not the man from Fenton (clue 1). The man from Fenton is Mr. Jacobs, the third gift-giver, and the man from Midway is first. By elimination, the Midway man is Mr. Olson, and the wagon was Ms. Miller's gift. Then the first three gifts were the water wheel, the Fulton woman's wagon, and the harness (clue 6). Ms. Alcott's household donation (clue 2) must have been the sewing machine and she was from Keeton (clue 3). By elimination, Mr. Reyes is from Canton. In summary, in order of donation:

> Mr. Olson, Midway, water wheel
> Ms. Miller, Fulton, wagon
> Mr. Jacobs, Fenton, harness
> Ms. Beebe, Senton, trunk
> Mr. Reyes, Canton, stove
> Ms. Alcott, Keeton, sewing machine

25. ADDLED AIRPORT

According to the introduction, there were only two pairs of travelers. Georgette was half of the pair who flew on the Friendly Bird flight (clue 2). By clue 7, one pair flew on Non-Skeddo and another pair flew on Flight #760. Therefore, Flight #760 could only be the Friendly Bird flight carrying Georgette. Georgette's sister, her companion on Flight #760, was not Ada (clue 4) or Mary (clue 7); it was Martha. These two did not depart first (clue 7). The other pair (on Non-Skeddo) was Fred and his companion (clue 6); they must have been the pair who left first at 5:15 (clue 5). Fred's companion was not Egbert (clue 2) or Mary (clue 7); Fred flew with Ada. The others flew in this order: Egbert flew after Georgette (clue 2) and Mary flew before them (clue 7). Therefore, Mary flew on Buy & Fly which left at 5:25 (clue 5) while by elimination, Egbert took the Icarus, Inc. flight. Flight #409 was not Egbert's (clue 2), nor could it be Fred and Ada's (clue 4); so it was Mary's. Flight #148 did not leave first (clue 3), so that was not Fred and Ada's flight; they were on Flight #211 and Egbert was on Flight #148. By clues 1 and 4, these two flights left from Concourses C and D. Egbert did not leave from C (clue 3), so Fred and Ada did, and Egbert left from D. Mary, who left at 5:25, did not leave from Concourse B (clue 6), she left from A and, by elimination, Georgette and Martha left from Concourse B. They left at 5:40 (clue 6) and Egbert left at 5:50 (clue 2). In sum:

> 5:15: Fred and Ada, Non-Skeddo Flight #211, Concourse C
> 5:25: Mary, Buy & Fly Flight #409, Concourse A
> 5:40: Georgette and Martha, Friendly Bird Flight #760, Concourse B
> 5:50: Egbert, Icarus, Inc. Flight #148, Concourse D

26. THE DEEP FREEZE

The warmest temperature was 0° (clue 4). By clue 1, then, the temperatures were 0°, −5°, −10°, −15°, and −20°. The student in the warmest town was not Bob (clue 4), Alice (clue 5), Ed (clue 7), or Carol (clue 8); he was Don. His last name is not Jackson (clue 4), Fry (clue 9), Hayes (clue 10), or Ives (clue 11); it is Goodman. The physics major's hometown had a temperature of −10° (clue 3). Ives was the art major (clue 11). Don Goodman's major is not history (clue 6) or math (clue 9); it is English. The history major is not Fry or Hayes (clue 6); he is Jackson. Fry is not the math major (clue 9); he is the physics major and, by elimination, Hayes is the math major. Alice is from Tennessee (clue 5). Illinois had the coldest temperature (clue 2). Don Goodman the English major is not from Iowa (clue 6) or Michigan (clue 7); he is from New York. Fry the physics major, whose hometown had a temperature of −10°, is not from Tennessee (clue 3) or Iowa (clue 6); he is from Michigan.

Hayes is not from Iowa nor, since her hometown was not the coldest (clue 6), was she from Illinois (clue 2); she is from Tennessee, and is Alice. By clue 6, Alice Hayes' hometown was −5°. By elimination, Iowa was −15°. Ed was not from Michigan (clue 7) or Illinois (clue 8); he was from Iowa. By clue 6, then, Jackson the history major went to Illinois; by elimination, Ed is Ives, the art major. Jackson is not Bob (clue 4); she is Carol and, by elimination, Bob is Fry. In summary:

> 0°: Don Goodman, English, NY
> −5°: Alice Hayes, math, Tenn.
> −10°: Bob Fry, physics, Mich.
> −15°: Ed Ives, art, Iowa
> −20°: Carol Jackson, history, Ill.

27. SOUTH OF THE BORDER

The Mexico residents are not the Cabots or the Moores (clue 1), the Nobles (clue 4), or the Tylers (clue 5); they are the Allens mentioned in clue 2. One tourist couple is childless (clue 7). By clue 1, the other three couples, one of whom drove a van, have children—including one girl and three boys—and are the Cabots, the couple from New Mexico, and the Moores; one of the last two are the Allens' friends. The Cabots and the Moores are (not necessarily respectively) from California and Texas, and each couple has one child (clue 3); the New Mexico couple then have two children and the childless couple are from the remaining state, Arizona. These are the Nobles, and the Tylers are from New Mexico (clue 5). Also by clue 5, since the Tylers didn't know the Allens before, the Moores did, and the daughterless couple with the van must be the Cabots, who thus have one son. The Moores' one child is also a son (clue 2), so the Tylers have a son and a daughter. Since the Cabots went by van, the Californians who went by train (clue 6) are the Moores, and the Cabots are from Texas. The Nobles and the Tylers, who chose Mexico because it was close to home, flew (clue 4). The Nobles chose Mexico for the sightseeing, the Cabots because it was inexpensive (clue 7). In sum:

> Allens, residents
> Cabots from Texas: van, inexpensive, one son
> Moores from California: train, friends, one son
> Nobles from Arizona: plane, sightseeing, no children
> Tylers from New Mexico: plane, close to home, one son and one daughter

28. MERRY-GO-ROUND RIDERS

Mary moved back only one position to her second mount, a zebra (clue 3), so she is neither of the two girls who were directly opposite one another and who exchanged mounts (clue 1). Mary is the third girl. Sue is one of the girls who exchanged mounts and she is opposite the Cain girl (clue 1), so Wilma is Cain. Mike's second mount was directly opposite mount #2 (clue 2), so he must have been on #5 for the second ride. Also, by clue 2, Mike was just behind Wilma, so she was on #4. Due to the exchange in clue 1, Wilma's first ride was #1, which was Sue's second mount; Sue's first mount was #4. Mary's first mount was numbered higher than Sue's #4, and Mary was in front of Mills (clue 3), so Mary was on #5 while Mills was on #6. By elimination, the first riders on both #2 and #3 were boys; the third boy was next to Mary, and he must be Mills. By clue 4, Ted's first mount was #2; Brown, a boy, was on #3, while Sue was on the llama. The zebra was mount #6 (clue 3). By elimination, Ted isn't Brown or Mills; he must be Mrs. Keyes' son. If Mike were Mills, he and Mary would have exchanged mounts for the second ride, contradicting clue 1, so Mike isn't Mills, he is Brown, and Mills is Greg. Each child rode a different mount for the second ride (introduction), so Ted's second mount wasn't #2; he was on #3. By elimination, Greg's second mount was #2. Mary isn't Mead (clue 5), so Sue is, while Mary must be Smith. Sue Mead didn't ride the lion, burro, or horse (clue 5); her second mount was the bear, which was Wilma's first ride. The burro's number was lower than the lion's, which was lower than the horse's (clue 5); therefore, mount #2 was the burro, #3 the lion, and #5 the horse. In summary, with the first mounts and numbers listed first:

142

Wilma Cain: #1 bear, llama
Ted Keyes: #2 burro, lion
Mike Brown: #3 lion, horse
Sue Mead: #4 llama, bear
Mary Smith: #5 horse, zebra
Greg Mills: #6 zebra, burro

29. KEEPING UP WITH THE JONESES

From clue 5, there are two sons—one named Sandy—and two daughters. The oldest is female, the second oldest a male horticulturist (clue 2). The youngest of the four is male (clue 7), so the second youngest is female. The younger brother is in advertising (clue 6) and is the only one who stayed in the city where he attended school (clue 7); that city is San Diego (clue 1). The younger sister is a management consultant, and Fran is her younger brother, while one of the two older siblings lives in San Francisco (clue 4). Sandy is then the older son. Lee lives in San Francisco (clue 6) and must therefore be the older sister. Also, by clue 6, the one who owns a hotel isn't Lee and must be Mrs. Jones, while Lee is the interior designer. The female sibling who lives in Sacramento (clue 1) must be the younger sister. Rusty, who lives in Pine Hill (clue 3) can only be Mrs. Jones. Be elimination, Sandy lives in San Mateo and the younger sister is Chris. In sum, with offspring listed in order from oldest to youngest:

Rusty (Mrs. Jones): hotel owner, Pine Hill
Daughter Lee: interior designer, San Francisco
Son Sandy: horticulturist, San Mateo
Daughter Chris: management consultant, Sacramento
Son Fran: advertising, San Diego

30. ON CAMPUS

We are told the four room numbers are 402, 409, 435, and 436, and two women and three men are mentioned. By clue 7, the two women occupy the first two rooms and the last two are occupied by the three men, two of whom are roommates. The roommates are in 435, and the man in 436 is the math major (clue 5). The woman in 402, who doesn't date, is Young (clue 2). By clue 4, then, the two students in 435 are the one from California and the biology major, the math major is from Wyoming, and Susan lives in 409; Young's first name is Karen. Jack is from California and Susan is from Pennsylvania (clue 6). We know the one in 436 doesn't know any of the others. By clue 1, then, Peter is Jack's roommate the biology major, and Karen Young's major is chemistry. The math major, by elimination, is Bill. Since clue 3 mentions all three men, Peter is from Maine, and Jack's last name is Pierce—and since he doesn't major in history, Susan does, and Jack's major, by elimination, is psychology. Peter's surname is Horne (clue 5). Karen Young, by elimination, is from Georgia. Bill's last name is Reilly (clue 4) and Susan's is Crane. In sum:

402: Karen Young, chemistry, Georgia
409: Susan Crane, history, Pennsylvania
435: Peter Horne, biology, Maine, and Jack Pierce, psychology, California
436: Bill Reilly, math, Wyoming

31. BUNKMATES

Ted has a bottom bunk (clue 8), so by clue 2, the brown blanket is on the top bunk over him. The green blanket (clue 3) and the tan blanket (clue 7) are also on top bunks, so the bottom bunks have gray, blue, and red, with the red on the right (clue 5). By clue 2, Ted is not at the right with the red blanket. Since Bill has the bottom center bunk (clue 5), the left bottom bunk is Ted's and the top left bunk has the brown blanket (clue 2). By clue 5, Ted is the Banks boy. By clue 9, then, the center bunk has the green blanket on top and the gray on the bottom. Ted, by elimination, has the blue blanket and the top right bunk has the tan one. By clue 7, the owner of the tan blanket is from Pennsylvania and below him is the boy from New Jersey. By clue 6, the bunk that has West on the bottom and the Connecticut boy on the top must be the center bunk. By clue 1, then, Bob has the top bunk on the left over Ted. To their right, by clue 2, must be Greer and the boy from Massachusetts, Greer from Connecticut on top and West from Massachusetts on the bottom. Ted, then, by clue 4, must be the boy from Pennsylvania with the bottom bunk. By elimination, Bob is from New York. Tim has a top bunk but not a green blanket (clue 3), so he is the other boy from Pennsylvania in the bunk with the tan blanket. By clue 6, then, Paul is below him and Tim's last name is Byrd. By elimination, Dave has the green blanket top center. Paul from New Jersey isn't named Winston (clue 7); by elimination he's the Lake boy and Bob is Winston. In sum:

Top left: Bob Winston, NY, brown
Bottom left: Ted Banks, PA, blue
Top center: Dave Greer, CT, green
Bottom center: Bill West, MA, gray
Top right: Tim Byrd, PA, tan
Bottom right: Paul Lake, NJ, red

32. SUMMER ROMANCES

If Bill had fallen in love with Joan, then, by clue 4, they would have fallen in love with each other. But that contradicts clue 2 (remember that each of the six fell in love with just one of the other five). So Bill fell in love with either Betty or Sue. If the object of his affections were Betty, then, by clues 3 and 4, Bill fell in love with Betty, who fell in love with Mike, who fell in love with Joan—who, by elimination, fell in love with Jason, who, by clue 2, fell in love with Sue. But that contradicts clue 1. So we must conclude that Bill fell in love with Sue. Joan then fell in love with Bill (clue 2). Since Betty didn't fall in love with Jason (clue 3), Sue did. Jason, in turn, fell in love with Joan (clue 4). By elimination, Betty and Mike fell in love with each other. In sum:

Betty: Mike   Joan: Bill
Bill: Sue   Mike: Betty
Jason: Joan   Sue: Jason

33. BUSINESS PARTNERS

From clue 3, the five couples are Norma and her husband, George and his wife, the Abbots, Evans' partner and his wife, and Marie and her husband. Norma's and Marie's husbands, in one order or the other, are Harry and Ivan, while the remaining two men are John and Kevin (clue 5). Baker's partner is none of those four (clue 2), so he is George. Norma's husband has a partner (clue 4), as does Abbot (clue 7), so Marie's husband does not. Since Harry has a partner (clue 1), he is Norma's husband and Ivan is Marie's. To find out who is partners with whom we must go back to clue 3: the first four people mentioned there are involved in a partnership. The fourth one mentioned in that clue is Evans' partner, George is Baker's partner; Abbot can't be his own partner, so Harry is Abbot's partner. Now, since the first and third people mentioned are partners, the second and fourth are also partners, which means that George must be Evans' and the fourth man is Baker. Abbot's wife is not Paula (clue 1), or Rose (clue 6); she is Susan. Her husband is Kevin, and John's last name is Baker (clue 6). Kevin's partner Harry's last name is not Davis (clue 1), so it is Carney, and Ivan and Marie are the Davises (clue 1). George's wife is Rose, and John's is Paula (clue 6). In sum:

Kevin and Susan Abbot
John and Paula Baker
Harry and Norma Carney
Ivan and Marie Davis
George and Rose Evans

The partners: Abbot & Carney, Baker & Evans.

34. BUSH BOGGLE

Quigley did not plant the mock orange (clue 4), the azalea (clue 6), or the forsythia (clue 9), so Quigley planted either the lilac or the rhododendron. By clues 9 and 6, the Quigleys do not live at an end house, so they didn't plant lilacs (clue 1); they planted the rhododendron. Vince didn't plant the rhododendron (clue 2), so he is not Quigley. Vince doesn't live on a corner (clue 2), so he is not Pringle (clue 1), nor is he Olcott (clue 2), or McDonald (clue 5); he is Newcomb. The Quigleys with their rhododendron live on one side of Vince Newcomb and the Olcotts on the other (clue 2). The azaleas grow one house away from Vince Newcomb (clue 8) right next to the Quigleys (clue 6) and the house right next to that contains Doris and her husband (also clue 6). Thus, Doris and her husband and the Olcotts occupy the end houses, so the Quigleys occupy 824 at the center. According to clue 1, then, Doris is Mrs. Pringle and the Olcotts grow lilacs. By elimination, the McDonalds grow the azaleas. Since the azaleas are planted on one side of the Quigleys (clue 6), Vince Newcomb grows forsythia (clue 9). By clue 3, the only house that Carol can occupy is the McDonald's house and Todd is Mr. Quigley. Since Edith's house number is less than Quigley's (clue 4), then, Carol's and Doris' must be higher; Carol lives at 826 and Doris at 828. Carol McDonald's neighbor Betty (clue 5) is Mrs. Quigley who lives at 824. By clue 7, Alice is Mrs. Olcott and lives at 820 and Zack is Mr. McDonald. Steve does not live at 820 (clue 1), so he is Mr. Pringle and Wayne lives at 820. By elimination, Edith is married to Vince and the mock orange grows at the Pringles' house. In sum:

> Alice and Wayne Olcott, lilac, 820
> Edith and Vince Newcomb, forsythia, 822
> Betty and Todd Quigley, rhododendron, 824
> Carol and Zack McDonald, azalea, 826
> Doris and Steve Pringle, mock orange, 828

35. DENTIST APPOINTMENTS

Helen has only two daughters and two granddaughters (introduction). Since Marie, her cousin, her mother (clue 2), and her aunt (clue 6) are among the above-mentioned five, Marie's aunt and Marie's mother must be sisters, i.e., Helen's daughters, and Marie and her cousin are Helen's granddaughters who are married (introduction) with no children (clue 3). Marie and her cousin are Ms. Bradford and Beth respectively (also clue 3). From clue 2, Marie's aunt must be Ms. Harris, who had lunch with grandma Helen at 12 noon. Dentist appointments were at 10 a.m., 11 a.m., 12 noon, 1 p.m., and 2 p.m. (clue 1). Ms. Harris had an 11 a.m. dentist appointment (clue 6); Valerie's was after 11 a.m. (clue 5), so Valerie is Marie's mother and Ms. Harris is Christine. Beth, Valerie, and Marie, then, had lunch at 1 p.m.; Christine Harris and her mother Helen, at 12 noon (clue 2). The woman with the 12 noon appointment has one child, so isn't Helen (introduction), Beth, or Marie Bradford (clue 3). Valerie's appointment was at 12 noon, and Ms. Vaughn's at 10 a.m. (clue 5). Since Marie and Beth ate lunch at 1 p.m. (clue 2), neither had the 1 p.m. appointment; Helen did. Helen, then, cannot be Vaughn, so Beth is Vaughn, and Marie Bradford's appointment, by elimination, was at 2 p.m. Ms. Miller's appointment with Dr. Stevens was not at 12 noon (clue 4), so Valerie isn't Miller. She is Carter and Helen is Miller. Only Helen has two children, so she bought shoes (clue 3). From clue 4, neither Christine nor Helen Miller, who both saw Dr. Stevens, nor Valerie Carter, who saw Dr. Perkins, bought the skirt, nor did Beth Vaughn (clue 5). Marie Bradford bought the skirt and saw Dr. Perkins (clue 4). Beth bought the dress and saw Dr. Stevens (clue 4). Valerie didn't buy the umbrella (clue 5), so Christine did, and Valerie, by elimination, bought the blouse. In summary:

> 10 a.m.: Beth Vaughn, Dr. Stevens, dress
> 11 a.m.: Christine Harris, Dr. Stevens, umbrella
> 12 noon: Valerie Carter, Dr. Perkins, blouse
> 1 p.m.: Helen Miller, Dr. Stevens, shoes
> 2 p.m.: Marie Bradford, Dr. Perkins, skirt

36. PET VISITORS

By clue 3, Dave brings a man's favorite on Mondays. By clue 5, a woman's favorite comes Wednesdays. No pets come Sundays (clue 9). All this being true, by clue 6, Miss Hopper's favorite can only come on Tuesdays, Mrs. Jacob's on Fridays and a man's on Saturdays. By elimination, the other day with no pets scheduled is Thursday and the woman who likes Wednesday's pet the best is Mrs. Swift. Cathy's pet (clue 4) comes Saturdays and Mr. Stuart's on Mondays. Mr. Murphy's favorite, by elimination, is Saturday's pet. Since only on Monday, Tuesday, and Wednesday do three pets visit consecutively, by clue 1, the cat must come Mondays, Pedro Tuesdays, and Holly's pet Wednesdays. Frisky, by clue 2, is Dave's Monday cat. Sam, by clue 7, doesn't bring Friday's pet; he brings Tuesday's and Peg, by elimination, brings her pet Fridays. We've established the women's favorites come on Tuesday, Wednesday, and Friday and that Tuesday's pet is Pedro. By clue 5, then, Pedro must be the parrot and Pip Friday's pet. Since Buddy visits the day after the rabbit (clue 8), Buddy must come Saturdays and the rabbit Fridays. Fancy, then, must be the Wednesday pet by elimination. By clue 7, Fancy isn't the monkey; by elimination, Fancy's the retriever and Buddy is the monkey. In sum:

> Mon.: Dave, Frisky, cat, Mr. Stuart
> Tues.: Sam, Pedro, parrot, Miss Hopper
> Wed.: Holly, Fancy, retriever, Mrs. Swift
> Fri.: Peg, Pip, rabbit, Mrs. Jacob
> Sat.: Cathy, Buddy, monkey, Mr. Murphy

37. SCIENCE ASSIGNMENTS

By clue 1, the student who wrote about comets was the one who discussed Saturn. By clue 2, the first speaker, who was assigned Mercury, and the last speaker, who was assigned Pluto, wrote on topics with the same initial. The written topics can't be comets and constellations, so they can only be geodes and geysers. Sally wrote about geodes (clue 3), and her talk wasn't last (clue 4), so the one who wrote about geysers discussed Pluto, and Sally discussed Mercury. By clue 5, two of the four girl students reporting discussed planets with their own initials. We know Sally did not. Erica spoke after all three of the other girls (again, clue 4), so she didn't give the report on Earth, which was third. Clue 5 must mean that Jill discussed Jupiter and Mary discussed Mars. From clue 6, then, Mary, who discussed Mars, did not write about meteorites. However, by clue 7, one student's planet and writing topic did share an initial. The only planets left that fulfill this clue are Venus and Earth. It can't be Venus and volcanoes, since Venus is second and the writer on volcanoes spoke at least third, after both the writer on constellations (clue 8) and Sally, first, on geodes; so the speaker on planet Earth wrote about eclipses. Then Earl's paper wasn't on eclipses (clue 6). By clue 9, Mark's wasn't on meteorites. The two boys mentioned in clue 10 whose name and writing topic shared an initial can only be Gavin on geysers and Tom on tides. Geysers, we know, were with Pluto, so that was Gavin's planet and from clue 10, Tom's was Uranus. By clue 11, John spoke just after a girl who wrote about satellites. She isn't Sally, who wrote about geodes, nor can she be Mary, who spoke just before Jill. Erica, the last girl speaker, must have given her report on either Saturn or Neptune, just before either Tom or Gavin. Only Jill can be the satellites writer, so John spoke on Saturn. Erica, then, spoke on Neptune. From clue 8, the papers on constellations and volcanoes must have been those of the second and fourth speakers, respectively, while Erica's topic was, by elimination, meteorites. Earl didn't discuss Earth and write about eclipses (clue 6), so Mark did, and Earl must have discussed Venus. In sum:

> Mercury, geodes, Sally
> Venus, constellations, Earl
> Earth, eclipses, Mark
> Mars, volcanoes, Mary
> Jupiter, satellites, Jill
> Saturn, comets, John
> Uranus, tides, Tom
> Neptune, meteorites, Erica
> Pluto, geysers, Gavin

38. SUPER-DUPER SANDWICHES

The least-expensive sandwich was not bought by Owens (clue 1), Kane (clue 4), or Miller (clue 6); so it was bought by either Nix or Lemon. Bill is neither Nix nor Lemon (clue 2), so he didn't buy the least expensive sandwich; nor did Earl (clue 1), Dick (clue 5), or Andy (clue 6), so Chad did. His last name is not Lemon (clue 8), so it is Nix. Andy is not Owens (clues 1, 6), Miller (clue 6), or Lemon (clue 8), so he is Kane. Chad then had the "Diamond Jim Brady" (clues 4, 6). Andy didn't buy the "Wide Load" (clues 1, 6), "Eiffel Tower" (clue 3), or "Belt Buster" (clue 4), so he bought the "Stomach Stunner." Neither Owens (clue 1) nor Miller (clue 9) had the "Wide Load," so Lemon did. His first name isn't Earl (clue 1) or Bill (clue 2), so it is Dick. Earl isn't Owens (clue 1), so he is Miller and Bill is Owens (clue 1). If Earl Miller had bought the "Belt Buster," then he would have spent $2.00 more than Andy (clue 4) and also twice as much as Andy (clue 6)—meaning Earl would have spent $4.00, Andy $2.00, Chad $1.00 (clue 6), Bill $3.00, and Dick $1.50 (clue 1)—a total of $11.50, conflicting with clue 7. Therefore, Earl bought the "Eiffel Tower" and Bill Owens the "Belt Buster." Using the price Chad paid as a basis for pricing the other sandwiches (let what Chad paid equal a ?), we know that Andy spent twice that amount ($2 \times ?$, clue 6), Earl paid four times that amount ($4 \times ?$, also clue 6), Bill paid the same as Andy plus $2 ($2 \times ? + \$2$, clue 4) and Dick paid half of that ($1 \times ? + \$1$, clue 1). When you add them all up, it comes to 10 times some amount plus $3 equals $18; the amount has to be $1.50. Therefore, Chad's sandwich cost $1.50; Andy's, $3.00; Earl's, $6.00; Bill's, $5.00; and Dick's $2.50. In sum:

> Earl Miller, "Eiffel Tower," $6.00
> Bill Owens, "Belt Buster," $5.00
> Andy Kane, "Stomach Stunner," $3.00
> Dick Lemon, "Wide Load," $2.50
> Chad Nix, "Diamond Jim Brady," $1.50

39. ROCK COLLECTIONS

Each child brought in 12 rocks to total 48 (introduction); 12 were white, 12 red, and 12 green (clue 1), so the other 12 were black. Lucy, who had twice as many green rocks as Terry (clue 7), had 2, 4, or 6 (clue 2); she had three times as many as Cory (clue 7), so Lucy had 6, Terry had 3, and Cory had 2 green rocks to total 11. Jody, then, had 1 green rock to total 12. Lucy had twice as many red rocks as white rocks (clue 4); since she had 6 green rocks and at least 1 black rock (clue 2), she could have no more than 5 white and red rocks. Since 4 red plus 2 white rocks would total 6, Lucy had 2 red rocks and 1 white rock, plus 3 black rocks to total 12. Terry had 3 green and 3 white rocks (clue 5). Jody had three times as many white rocks as black rocks (clue 6), so he had 3 or 6 white rocks (clue 2) and 1 or 2 black rocks. If he had 3 white rocks and 1 black rock, plus his 1 green rock, he'd need 7 red rocks to total 12, but clue 2 limits him to 6 red rocks, so Jody had 6 white, 2 black, 1 green, plus 3 red rocks to total 12. With Jody's 6 white rocks, Lucy's 1, and Terry's 3, Cory had 2 white rocks to total 12. Cory had at least 3 red rocks (clue 3), and Terry had at least 4 red rocks (clue 5) to total 7. With Lucy's 2 red rocks and Jody's 3 red rocks, that accounts for all 12 red rocks (clue 1). Cory, then, had 5 black rocks to total 12, and Terry had 2 black rocks to total 12. In summary:

> Cory: 5 black, 2 green, 3 red, 2 white
> Jody: 2 black, 1 green, 3 red, 6 white
> Lucy: 3 black, 6 green, 2 red, 1 white
> Terry: 2 black, 3 green, 4 red, 3 white

40. FUND-RAISING EVENTS

From the introduction, the five Saturdays in question were June 2, 9, 16, 23, and 30. June 2 was the day that Faith Baptist Church held its fund raiser (clue 4). By clue 9 and the fact that the fifth Saturday's event was held on June 30, Wister headed the Faith Baptist Church's committee, Trinity United Methodist Church's event was June 9, and the spaghetti dinner was on June 23. Neither the spaghetti dinner (clue 1) nor the event put on by the United Methodists (clue 8) was planned by Forrest's committee, so that committee planned either June 16's fund raiser or June 30's fund raiser. Suppose the latter were the true case. By clue 8, Alan and his committee would have put on the event of June 23. By clue 3, Mr. Browder's committee's fund raiser would have been on June 9, Joe's on June 16, and Nichols would be Alan's last name. By elimination, the only man left, Bill, would be Mr. Browder. Clue 7, then, could not be satisfied. Therefore, Forrest's committee planned June 16's fund raiser, and Alan headed the committee that planned the Methodist event. By clue 3, Alan is Browder. Also by clue 3, Joe's committee planned the event of June 16 or June 23; by clue 7 Jean's event was also on either of these two dates. Since the fund raiser headed by Jean was on the Saturday preceding the one put on by the Unitarian Church (clue 7), it can't have been on June 16, as that would mean Joe's event would have been on June 23 (clue 3), making Joe a member of the Unitarian Church, which contradicts clue 10. Thus, the fund raiser headed by Joe was on June 16 and he is Forrest, Jean's event was on June 23, the fish fry was on June 16, and the event put on by the First Unitarian Church was on June 30. Bill isn't Wister (clue 5), so he headed the planning of June 30's fund raiser; by elimination, Mary is Wister. Jean isn't Nichols (clue 2), so Bill is, and Woods can only be Jean (clue 11). By clue 5, the barbecued chicken dinner was held on June 9. By clue 6, the pancakes-and-sausage breakfast was not held on June 30, it was held on June 2. Thus, by elimination, the crab picking feast was held on June 30. By clue 11, Woods's church isn't St. Mary's Catholic, so it is St. Luke's Episcopal, and Forrest's fish fry was put on by St. Mary's Catholic Church. In summary:

June 2: Faith Baptist Church, pancakes-and-sausage breakfast, Mary Wister
June 9: Trinity United Methodist, barbecued chicken dinner, Alan Browder
June 16: St. Mary's Catholic Church, fish fry, Joe Forrest
June 23: St. Luke's Episcopal Church, spaghetti dinner, Jean Woods
June 30: First Unitarian Church, crab picking feast, Bill Nichols

41. WEEKLY CHORES

Clue 1 lists four tenants who washed clothes on consecutive days, as does clue 5. Clearly, these lists are not identical—but they must overlap, else there would be eight days described. Further, they must overlap by two or three, since an overlap of just one would give us seven days. There are just two possibilities that do not result in conflicting first or last names. One is that the Monday-through-Saturday sequence was: Missy, one who vacuumed Saturday, Wanda Lange, Laura Hogan, Flora, and one who vacuumed Monday. But there is no place in this sequence for Susan to have done laundry the day before Ms. Bates (clue 7). Therefore the correct sequence, one covering five days, is: Wanda; Missy Hogan; Flora, who vacuumed on Saturday; Ms. Lange, who vacuumed on Monday; and Laura. By clues 2 and 7, neither Vicki nor Susan washed clothes on Saturday, so Laura did. Wanda's day was then Tuesday, Missy Hogan's Wednesday, Flora's Thursday, and Ms. Lange's Friday. By clue 7, then, Susan can only have done laundry on Monday, Wanda's last name is Bates, and Missy Hogan vacuumed on Monday. Ms. Lange, by elimination, is Vicki; Laura is Ms. Magers (clue 2). Flora is Ms. Bookman, and Missy Hogan's second vacuuming day was Friday (clue 4). Susan, by elimination, is Ms. Gleason. By clue 9, Susan Gleason and Laura Magers vacuumed Tuesday through Friday. Since no one vacuumed on consecutive days (clue 6), one vacuumed Tuesday and Thursday, the other Wednesday and Friday. The former is Laura, the latter Susan (clue 3). We know that Monday, Ms. Lange vacuumed; Tuesday, Ms. Bates did laundry; and Friday, Ms. Lange did laundry. By clue 8, one of them vacuumed on each of the other three days. Vicki Lange didn't vacuum on Wednesday (clue 2), and Ms. Bates cannot have vacuumed both Wednesday and Thursday (clue 6). Therefore, Ms. Lange vacuumed Thursday, Ms. Bates Wednesday and Saturday. Flora's second vacuuming day was Tuesday (clue 10). In sum, with vacuuming days in parentheses:

Mon.: Susan Gleason (Wed., Fri.)
Tue.: Wanda Bates (Wed., Sat.)
Wed.: Missy Hogan (Mon., Fri.)
Thurs.: Flora Bookman (Tue., Sat.)
Fri.: Vicki Lange (Mon., Thurs.)
Sat.: Laura Magers (Tue., Thurs.)

42. BABY-SITTERS

Clue 1 lists three women who baby-sat on consecutive evenings, as does clue 2. They are not the very same women, since the third in clue 1 sat with a boy and the third in clue 2 with Linda. Thus, either they are six distinct women, or there is some degree of "overlap" between the two lists. Mrs. Usher sat with a girl (clue 7) so she is not the third woman mentioned in clue 1. Helen sat with Nancy (clue 5), so she is not the third woman mentioned in clue 2. Thus, there is no "overlap," and all six women are mentioned in the two clues. The three evenings mentioned in clue 1, since the woman who sat with the Timmons boy did not do so on Saturday, were Monday through Wednesday, those in clue 2 Thursday through Saturday. Therefore, putting the two together, Helen sat with Nancy Monday, Mrs. Wilson baby-sat Tuesday, the Timmons boy was left with one of the women Wednesday, Mrs. Usher kept one of the girls Thursday, Gloria baby-sat Friday, and one of the woman sat with Linda Saturday. By clue 7, Mrs. Usher's charge must have been Julie. By clue 3, Doris baby-sat on Tuesday, Mrs. Rivers on Wednesday, and Gloria kept Marty. Helen is Mrs. Timmons, Doris Wilson kept Kevin (clue 9), and, by elimination, Ian is the Timmons boy. Gloria isn't Mrs. Short (clue 3), she is Mrs. Vance, so Linda's baby-sitter was Mrs. Short (clue 3). Faith is Mrs. Rivers, and Linda is the Usher girl (clue 10). Ellen isn't Mrs Usher (clue 6), so she is Mrs. Short, and Carla is Mrs. Usher (clue 6). Nancy's last name is Vance (clue 8). Kevin's last name isn't Wilson or Rivers (clue 4), it is Short. Marty's is Rivers (clue 11), and Julie is Wilson. In sum:

>Mon.: Helen Timmons kept Nancy Vance
>Tue.: Doris Wilson kept Kevin Short
>Wed.: Faith Rivers kept Ian Timmons
>Thu.: Carla Usher kept Julie Wilson
>Fri.: Gloria Vance kept Marty Rivers
>Sat.: Ellen Short kept Linda Usher

43. HOW OLD IS YOUR CAT?

Each family has two children and a cat. Ted and a brother are in one family and Nadine and a sister are in another (clue 3); Sam and a sister are in a third family (clue 4), so the fourth family has a boy and a girl. Ted's brother is Rodney (clue 5). Since no child is the same age as the family's cat (clue 8), Sam is not Carrie's 10-year-old brother (clue 6); Brad is, and they are the fourth family. The oldest child is 12 (clue 7). For Ted to be twice Tinker Bell's age, who is twice Tanya's age (clue 1), Ted must be 12, 8, or 4. Ted is four years older than Rodney (clue 5). Nadine's sister is also four years older than Rodney and Nadine is four years older than her sister (clue 3). Therefore, Nadine is 12, her sister and Ted are 8, and Rodney is 4. By clue 1, then, Tinker Bell is 4, and Tanya is 2. Tanya, then, is not Nadine's sister; Julie is, and Tanya is Sam's sister. Since Julie is 8, so is Fuzzy (clue 2). Pixie and Tanya are both 2 (clue 4) and Spooky is 6 (clue 5). Sam's age is the same as that of Carrie's cat (clue 6); since all cats' ages are even numbers, Carrie is the 7-year-old. Julie's cat is half as old as Julie (clue 2), so it is the 4-year-old Tinker Bell. In one family, the cat is older than the children, and in three families, the cat is younger than both children (clue 8). Ted and Rodney, at 8 and 4, are not both younger than any cat, so they must be older. Their cat, then, is 2-year-old Pixie. Brad and Carrie, at 10 and 7, are not both younger than any cat, so they are both older, and their cat is 6-year-old Spooky. Sam is the same age as Carrie's Spooky (clue 6), so he is 6, and he and Tanya live with 8-year-old Fuzzy, who is older than both. In summary:

>Brad (10), Carrie (7) & Spooky (6)
>Julie (8), Nadine (12) & Tinker Bell (4)
>Rodney (4), Ted (8) & Pixie (2)
>Sam (6), Tanya (2) & Fuzzy (8)

44. DIET CHALLENGE

We know that the dieters' ages can range from thirty to thirty-nine, and only one is an odd number. The thirty-two-year-old is not the youngest (clue 2), so the youngest dieter is either thirty or thirty-one. If the latter, another dieter would be thirty-eight (clue 5)—and from clue 1, the fourth dieter would be thirty-five, a second odd number. Therefore, the youngest dieter must be thirty; he or she started at 170 pounds, and a thirty-seven-year-old lost five pounds (clue 5). The fourth dieter, from clue 1, is thirty-four and initially weighed 180 pounds. Diane lost fourteen pounds, and the clerk lost twelve pounds (clue 3). We know the fourth person lost sixteen pounds (clue 3). Since the lightest dieter initially weighed 150 pounds (introduction), only the sixteen pound loss can represent a loss of 10 percent (clue 4). Therefore, by clue 4, the accountant initially weighed 160, lost sixteen pounds, and achieved a final weight of 144 pounds; this person can only be the thirty-two-year-old. By elimination, the thirty-seven-year-old initially weighed 150 pounds and, with the five-pound loss, ended the diet weighing 145. Diane lost fourteen pounds (clue 3), so she's not the thirty-seven-year-old, or the 180-pound, thirty-four-year-old (clue 1), or the thirty-two-year-old accountant who lost sixteen pounds; she is the thirty-year-old and weighed 170 pounds initially, with a final weight of 156 pounds, so the clerk, who lost twelve pounds, initially weighed 180 pounds, with a final weight of 168 pounds (clue 3). Alex is the one who lost five pounds (clue 4). He is not the typist (clue 1), so he is the sales representative, and Diane is the typist. Mary is not the clerk (clue 3), so John is, and Mary is the accountant. In sum:

Alex, 37, sales rep, lost 5 lbs. to weigh 145
John, 34, clerk, lost 12 lbs. to weigh 168
Mary, 32, accountant, lost 16 lbs. to weigh 144
Diane, 30, typist, lost 14 lbs. to weigh 156

45. LOST IS FOUND

The last names of the five women are Castle (clue 2), King (clue 1), and Bishop (clue 3). Mrs. King sent the first item—a letter to her sister (introduction, clue 1). Of the other four, Joyce and Mrs. Castle each sent something to a daughter (clue 2). A third sent something to a grandmother, and the other sent a textbook to a sister (clue 5). Joyce is not Castle or King (clue 2) or one of the Bishop twins (clues 2, 3) so she is Mother Bishop. Paula is not one of the Bishop twins (clue 3) and, since she has no children, she is not Mrs. Castle (clue 2). Paula is Mrs. King. Helen has no sisters, so is not one of the Bishop twins (clue 3). Helen is Castle, and the Bishop twins are Karen and Christine—one of whom, by elimination, sent a book to a sister and the other something to their grandmother. From clue 1, each person except Paula King received something before sending something, so each recipient must be the next in line to have sent something, with Paula being the last recipient. Since the twins' last name is Bishop, Helen Castle sent nothing to Christine or Karen (clue 2) or Joyce (clue 6); she sent the last item to daughter Paula, and was thus the fourth to receive something (clue 1), but not from Joyce (clue 6) or Karen, who received and sent something before Christine received or sent anything (clue 4). Thus Christine sent the fourth item to Helen, who is not her sister (clue 3), so must be her grandmother; it was not the photos (clue 2), the sunglasses (clue 4), or the book (clue 5). Christine sent the sweater pattern and received the third item—a book—from her sister, Karen (clue 5). Karen, then, was the second to receive something and, by elimination, Joyce the second to send something. Since Helen received the pattern, Joyce sent the photos to daughter Karen (clue 2). By elimination, Joyce received the letter from sister Paula, and Helen sent the sunglasses. In summary:

1. Paula King sent letter to sister Joyce
2. Joyce Bishop sent photos to daughter Karen
3. Karen Bishop sent book to sister Christine
4. Christine Bishop sent sweater pattern to grandmother Helen
5. Helen Castle sent sunglasses to daughter Paula

46. MR. TAXI'S FARES

According to the introduction there were five rides all together. Three of the four rides mentioned in clue 1 were Ms. Becker's, the one starting from Briar Lane, and the one to the zoo. Carl was picked up on Lexington Avenue and did not go to the zoo (clue 11), so he was a fourth fare. The one picked up on Beacon Drive did not go to the zoo (clue 4) and was not Ms. Becker (clue 7), and so was the fifth customer. Since Mahoney was picked up on

Broadway (clue 8), he or she had to be the one who went to the zoo. Ms. Becker, by elimination, was picked up on Maine Street. By clue 3, neither Ms. Parker nor Sam was picked up on Briar Lane, so Ms. Parker was picked up on Beacon Drive, and Sam can only be Mahoney. Ms. Becker, picked up on Maine Street, is neither Alice (clue 1) nor Eileen (clue 10): she is Debra. Smith was not picked up on Lexington and is not Eileen (clue 6); she must be Alice and was the one picked up on Briar Lane. By elimination, Carl's last name is Haley and Ms. Parker's first name is Eileen. By clue 1, neither Debra Becker nor Alice Smith, who was picked up on Briar Lane, went to the theater; nor did Carl Haley (clue 5), so Eileen Parker did. Debra Becker did not go to the bank (clue 7) or the airport (clue 9) and must have gone to the mall. Alice's ride cost $10 more than Debra's (clue 2), while the ride to the bank was $4 less (clue 7), so Alice cannot have gone to the bank and must have gone to the airport, while Carl went to the bank. Debra's ride cost $10 less than Alice's ride to the airport (clue 2) which was half of Alice's fare (clue 9), so Alice's ride to the airport cost $20, and Debra's ride cost $10. Then, Sam Mahoney's ride cost $15. By clue 7, Carl's ride to the bank cost $6, and Eileen's from Beacon Drive $12. In sum:

Debra Becker: from Maine St. to mall, $10
Carl Haley: from Lexington Ave. to bank, $6
Sam Mahoney: from Broadway to zoo, $15
Eileen Parker: from Beacon Dr. to theater, $12
Alice Smith: from Briar Lane to airport, $20

47. WHO HAD THE DAY OFF?

Three women are mentioned. Tina's day off was no later than Wednesday (clue 2). Molly's was no later than Thursday (clue 3), as was Fay's (clue 4). So those who were off Friday and Saturday were men. Will's day off was no later than Tuesday (clue 1), Ted's no later than Wednesday (clue 5). By clue 7, then, Sid was off Friday and Mr. Flint, who must be Sam, Saturday. Molly isn't married to Sid (clue 6) or Sam Flint (clue 7), so, by clue 3, we now know she was off Tuesday or before. We now know that Molly and Will were off Tuesday or before, and Ted and Tina were off Wednesday or before, so the one who was off Thursday had to be Fay. By clue 3, Fay isn't Mrs. White who was off two days before her husband, nor is Mrs. White Molly, so Tina is. Tina White wasn't off Tuesday or Wednesday (clues 3, 6), or Sunday (clue 8), so she was off Monday, and Mr. White, who must be Ted, was off Wednesday. By clue 3, then, Molly was off Sunday, and her husband Will was off Tuesday; they are the Marshes (clue 2). The third married couple, the Smiths, must then be Fay and Sid. In sum:

Sunday: Molly Marsh
Monday: Tina White
Tuesday: Will Marsh
Wednesday: Ted White
Thursday: Fay Smith
Friday: Sid Smith
Saturday: Sam Flint

48. NAME MATCH

We are told that Ida Lee is one of the five women. The one whose first name is Mary is the shortstop (clue 2). Sue is a middle name (clue 3). Since Lee and Sue are middle names, Lou and Sarah are first names (clue 1). By clue 7, the fifth first name is Anne, and another woman's middle name is Anne. By elimination, Jane and Marie are the remaining middle names. By clue 5, the one whose middle name is Jane, who plays first base, isn't the one whose first name is Lou; nor is her first name Anne, since, by clue 7, the one whose first name is Anne is either the catcher or plays something other than first base; so it is Sarah, and, by clue 6, her last name is Luce. The one whose middle name is Marie plays second base (clue 4), so her first name is either Anne or Lou. If she is Lou Marie, then she isn't Ms. Madison (clue 7), so she would not be one of the two cases mentioned in clue 8. There would be two middle names remaining to be assigned, Anne and Sue, so Mary would be Mary Anne, and Anne would be Anne Sue; by clue 8, their last names would have to be, respectively, Adams and Saunders. But by clue 7, one would have to be Ms. Madison, so this is impossible. Therefore, the first name of the one whose middle name is Marie is Anne. By clue 7, she is Ms. Madison, and the one whose middle name is Anne is the catcher; the latter can only be the one whose first name is Lou. By elimination, Mary's middle name is Sue, and Ida Lee plays third base. Anne Marie Madison is one of the two mentioned in clue 8. Mary Sue isn't Ms. Irving (clue 3). If she were Ms. Adams, then there could be only one woman fitting clue 8's description. She must be the second woman referred to in clue 8— i.e., her last name is Saunders—and, since there are exactly two such women, Lou Anne is not Ms. Adams and must be Ms. Irving, while Ida Lee is Ms. Adams. In sum:

Ida Lee Adams, 3rd base
Lou Anne Irving, catcher
Sarah Jane Luce, 1st base
Anne Marie Madison, 2nd base
Mary Sue Saunders, shortstop

49. MOWING LAWNS

Nick was paid $8, $9, $10, $11, and $12 (introduction). Frank didn't pay Nick the most or least (clue 2), nor did Joe (clue 4). There are only two women, so by clue 1 Nick mowed Sally's lawn on Wednesday; she did not pay him the most or least, and Mildred did not pay him the most. John must have paid Nick $12, Mildred $8. Nick mowed Hilton's lawn on Tuesday (clue 5). By clue 3, Miller paid Nick $10, and John's lawn was mowed either on Tuesday and Miller's on Wednesday, or John's was mowed on Thursday and Miller's on Friday. If Nick mowed John's on Thursday and Miller's on Friday, then by clue 2, Frank would have paid Nick $9 and Mildred Norris would have paid him $8 on Monday. Frank would have to be Hilton and Miller, by elimination, would be Joe. But this contradicts clue 4. So John is Hilton and paid Nick $12 on Tuesday and Sally Miller paid him $10 on Wednesday. Now, by clue 2, Nick must have been paid $11 on Friday, $9 for Frank's lawn on Thursday, and $8 for Mildred Norris's lawn on Monday. Nick mowed Joe's lawn Friday, and Frank is Carpenter (clue 4). Joe, by elimination, is Wilson. In sum:

Mon.: Mildred Norris, $8
Tue.: John Hilton, $12
Wed.: Sally Miller, $10
Thu.: Frank Carpenter, $9
Fri.: Joe Wilson, $11

50. MILL VALLEY PROGRAM

From clue 1, each grade provided a singer and an actor of opposite sex. From clue 2, the sixth-grade boy acted, so the sixth-grade girl must have sung. She can't be actor Ginny (clue 2) or actor Carol (clue 3), and as the oldest girl she can't be Debby or Sandie (clue 5) or Lisa (clue 8), so she must be Kate. From clue 7, then, sixth-grader Kate must be the singer in pink. The boy actor from her grade isn't Nick (clue 2), Paul (clue 4), Mark (clue 6), or John (clue 7), and as the oldest boy he can't be Bob (clue 8), so he must be Hank. His role wasn't Pinwheel, Nutmeg, or Carousel (clue 2), Pippin (clue 3), or Clove (clue 6), so it must have been Cinnamon. From clue 9, then, Clove and Nutmeg were also played by boys. Three of the six boys sang (clue 7), so only three can have acted. Nick acted, but not as Nutmeg (clue 2), so he must have played Clove. The girl in Mark's grade sang (clue 6), so Mark must have acted and must have played Nutmeg. The remaining actors are girls. Ginny acted, but not as Pinwheel or Carousel (clue 2), so she must have played Pippin. She isn't the youngest, or first-grade girl (clue 8), and actor Carol is even older (clue 3), so Carol can't be the second-grade girl who played Pinwheel (clue 2) and must be the one who played Carousel. Since the first-grade girl didn't act, she sang, and from clue 7 she is Sandie. The third-grade girl also sang (same clue), and we know sixth-grader Kate did, so actors Ginny and Carol can only be the girls from fourth and fifth grade, respectively. Since the girls from grades two, four, and five acted, the boys from those grades must be singers. Bob is in a lower grade than fourth-grader Ginny (clue 8), so he is the second grader and, from clue 7, he wore green. John's grade isn't fourth (same clue), so Paul's must be, and John's must be fifth. The girls who sang wore pink, lavender, and blue (clues 6, 7), so John and Paul in one order or the other must have worn white and yellow. The boy in yellow is one grade above the girl in blue (clue 10), but John is one grade above Paul, so Paul wore yellow and John white. The girl in blue, then, is the third grader who sang. Lisa is older than second-grader Bob (clue 8), so she is the girl in blue. Mark is in third grade with her (clue 6), so by elimination Nick is in first grade with Sandie, who must have worn lavender. By elimination also, Debby is the second grader who played Pinwheel. In summary:

Actors:
- Grade 1. Nick, Clove
- Grade 2. Debby, Pinwheel
- Grade 3. Mark, Nutmeg
- Grade 4. Ginny, Pippin
- Grade 5. Carol, Carousel
- Grade 6. Hank, Cinnamon

Singers:
- Grade 1. Sandie, lavender
- Grade 2. Bob, green
- Grade 3. Lisa, blue
- Grade 4. Paul, yellow
- Grade 5. John, white
- Grade 6. Kate, pink

51. KEEPING THE LIBRARY OPEN

Calling Monday a weekend day, we note that during the six days Saturday through Thursday, the volunteers each worked two weekend days and one weekday, while the librarians each worked one weekend day and two weekdays. On each weekend day, then, the library was staffed by two volunteers and one supervising librarian—while on each weekday, two librarians and one volunteer were present. There are three women and three men among the six; the women are Brooks (clue 1), Calhoun (clue 5), and Eastman (clue 7). Brooks had alternate days free and is not Helen (clue 1); nor is she Meg (clue 3), so she is Ann. She didn't work Saturday (clue 6), so she worked Sunday, Tuesday, and Thursday and is therefore a librarian. Helen, by clue 1, also worked alternate days; they were not the same days, since then she and Ann would both be librarians, and we know only one librarian was there on Sunday. Helen worked Saturday, Monday, and Wednesday, then, and is a volunteer. Ms. Calhoun is a librarian (clue 5), so she is Meg, and Helen's last name is Eastman. Meg Calhoun, who had consecutive free days (clue 3) must have worked Saturday, Wednesday, and Thursday. The second Saturday volunteer was Drake (clue 4). If Drake worked Monday with Helen, then only one volunteer would be left to work Sunday, so Drake worked Sunday. The third volunteer, Flint (clue 5), then worked Sunday and Monday. The remaining librarian, who worked Monday, Tuesday, and Wednesday, is Anders. By clue 2, Anders is John, and Flint, who worked Sunday and Monday, did not work Tuesday, but Thursday, while Drake worked Tuesday. Drake's first name is George and Flint's is Ned (clue 4). In sum:

Librarians: John Anders: Mon., Tues., Wed.
Ann Brooks: Sun., Tues., Thurs.
Meg Calhoun: Sat., Wed., Thurs.
Volunteers: George Drake: Sat., Sun., Tues.
Helen Eastman: Sat., Mon., Wed.
Ned Flint: Sun., Mon., Thurs.

52. DRIVING TESTS

Vane, the only one who received a combined score of less than a passing 150 (clue 1), must have scored 70 on each part of the test, for a total of 140. Everyone else scored at least 70 plus 80, so by clue 3, Jenny must have scored 80 on each part for a total of 160, while the White girl scored 150. Jenny, then, isn't the Taylor girl, who scored better on one part than the other, nor is her last name Root (clue 6) or Simms (clue 5), so it is Peters. The other two girls, Becky and Heidi, are the Taylor girl and the White girl, in one order or the other. Gary isn't Simms, and unlike Vane he passed (clues 2, 9), so he is the Root boy. Matt isn't Simms, either (clue 5), so he is Vane and Dave is Simms. We know that Matt Vane scored two 70s, Jenny Peters two 80s, and the White girl one 70 and one 80. The Taylor girl scored 80 on the quiz and 70 on the drive (clue 6). Gary Root scored 80 on the drive and either 70 or 80 on the quiz (also clue 6). Dave had at least one 80 in order to pass. There are at least six scores of 80, one for each of the five candidates who passed and an extra one for Jenny. Only Gary or Dave could have had an extra 80, but not both of them (clue 4), so at most there were seven scores of 80. It is impossible to allocate seven scores of 80 in such a way that two more 80s were awarded in the morning than in the afternoon (clue 7), so Gary and Dave each had just one, for a total of six 80s, four given in the morning and two in the afternoon. Gary Root scored 80 on the drive and 70 on the quiz, while Dave Simms' scores were the reverse (clue 4). Since the three afternoon candidates received only two 80s among them, one of the afternoon candidates must have been Matt Vane, the only one with no 80, and the other two candidates needed one 80 each to pass. Jenny, then, must have received her two 80s in the morning. The White girl was also a morning candidate (clue 3), but Heidi was an afternoon one (clue 8), so White is Becky's last name and Heidi's is Taylor. We know Becky and Jenny tested in the morning, Heidi and Matt in the afternoon, so Dave and Gary tested at opposite times. Morning scores on the quiz exceeded afternoon scores (clue 7). If Dave had tested in the afternoon, the quiz scores for the afternoon would have been Dave 80, Heidi Taylor 80, and Matt 70, for a total of 230; for the morning, they would have been Jenny 80, Gary 70, and Becky at most 80, totaling 230 at most. Therefore, Dave must have tested in the morning and Gary in the afternoon. Dave had only 70 on the practice drive, but there must be two 80's among the morning drive scores to exceed Gary's afternoon 80 (also clue 7), so Becky's 80 was on the drive, and she scored 70 on the quiz. In sum:

Morning	Drive	Quiz	Total
Jenny Peters	80	80	160
Dave Simms	70	80	150
Becky White	80	70	150
Total A.M. Scores	230	230	460

Afternoon	Drive	Quiz	Total
Gary Root	80	70	150
Heidi Taylor	70	80	150
Matt Vane	70	70	140
Total P.M. Scores	220	220	440

53. THE SHOPPERS

By the introduction, all six shopped on different floors, and each of the six went directly to her floor and made exactly one purchase. By clue 3, Ms. Patrick shopped on the third floor. By clues 4 and 6, Ms. Weeks bought her husband a jacket on either the first or second floor, while the washing machine was purchased three floors above—i.e., on either the fourth or the fifth. Ms. Thomas also shopped on either the first or second floor and didn't buy the crib (clue 8); by clue 5, then, the crib must have been bought by Ms. Patrick, and the lamp was purchased on the sixth floor. Ms. Thomas bought either the dress or the tablecloth. If the latter, it would be on the second floor (clue 2). The jacket would then have been purchased on the first floor, the washing machine on the fourth, and the dress, by elimination, on the fifth. But that contradicts clue 7. So Ms. Thomas bought the dress, by clue 7, on the second floor. Ms. Weeks then shopped on the first floor, and the washer purchase was made on the fourth; the tablecloth, by elimination, was bought on the fifth. By clue 2, then, Ms. Stern purchased the lamp. By clue 1, Ms. Vail can only have been the one who bought the tablecloth. Ms. Newton, by elimination, bought the washer. In sum:

6th: Stern, lamp
5th: Vail, tablecloth
4th: Newton, washing machine
3rd: Patrick, crib
2nd: Thomas, dress
1st: Weeks, jacket

54. SPRING BULB ORDERS

By clue 3, the second customer's order plus the daffodil order plus the Morris order equaled the number of dozens ordered of the bulbs to be planted by the front walk plus the only three-dozen order. We are told the largest order is no more than five dozen and the smallest no fewer than two dozen. Since there is only one three-dozen order (clue 3), then the only combination to satisfy clue 3 is two two-dozen orders plus one four-dozen order equaling a five-dozen order, plus the three-dozen order. By clue 1, then, the crocus order must have been for two dozen and the third order for four dozen. Since the third order was for four dozen, the second order, then, by clue 3, was for two dozen. The five-dozen order, by the same clue, was the one to be planted by the front walk. It wasn't the crocus (clue 1), hyacinth (clue 2), daffodil (clue 3), or jonquil (clue 4) order; it was, by elimination, tulips. The tulips weren't the second (clue 3), third (clue 1), fourth (clue 6), or fifth (clue 4) order; they were the first. By clue 4, the second order was to be planted around a mailbox. Therefore, the Stouts didn't order the tulips (clue 5), nor did the Merchants (clue 6), the Morrises (clue 3) or the Crocketts (clue 2); the Clevelands did. We've established that the first order was five dozen, the second two dozen, and the third four dozen. By clue 5, in order for the Stouts' order to be larger by one dozen than the one following, the fourth order had to be for three dozen and the fifth for two dozen. By clue 2, then, the Crocketts placed the third order for four dozen, the second order was for hyacinths, and the fourth, three-dozen order was for planting by a garage. The other two-dozen order, which was placed fifth, then, was for the crocuses (clue 1). By clue 3, the daffodils were the four dozen ordered third and the Morris family ordered the crocuses. By elimination, the jonquils were the three dozen ordered fourth. By clue 5, then, the Stouts placed the fourth order, and the fifth order was to be planted by the birdbath. By elimination, the third order of daffodils was to go by a front porch, and the Merchants ordered the hyacinths. In sum, as they ordered:

Cleveland, 5 doz. tulips, front walk
Merchant, 2 doz. hyacinths, mailbox
Crockett, 4 doz. daffodils, front porch
Stout, 3 doz. jonquils, garage
Morris, 2 doz. crocuses, birdbath

55. ALIEN LUV

The female crew members are from the planets Telme (introduction), Butduu, Lykta (clue 1), Weir (clue 2), and Conphes (clue 4). Therefore, the males hail from Som'i, Idluvit, Ovkorsnot, Sheram, and Chother. The male crew members' last names are Frendle (introduction), Ntel (clue 3), Withme (clue 5), Aleon (clue 8), and 4ee (clue 9). The females' last names, then, are Imuss, Yano, Ophens, Ther, and Uwood. The males' first names are Perfec (introduction), Goowt (clue 1), Iaman (clue 2), Ru (clue 3), and Kis (clue 6). The females' first names, then, are Hy, Nouh, Mab, Pherst, and Watta. The female from Conphes was married fourth (clue 4). By clue 6, Kis, Mab, and the male from Som'i were married in consecutive ceremonies—either second through fourth or third through fifth. Assuming the latter, by clue 1, Hy would have married second, the female from Butduu third, and Goowt fourth. But there would be no place, then, for the sequence described in clue 2. Therefore, Kis was married second, Mab third, and the male from Som'i fourth; the last then married the female from Conphes. Neither Goowt (clue 1) nor Iaman (clue 2) was married first or fifth, so they must be married third and fourth in one order or the other. By clue 3, Ru was not married fifth, so he can only be the male who married first, and Kis's last name is Ntel; Perfec was then the male married fifth. By clue 5, Withme, from Idluvit, was not the first or fifth; he can only be the male who married Mab. By clue 8, Nouh can only be the female who married second, Aleon was married fourth, and Yano was married fifth. Now, by clue 1, Hy married Ru, Nouh is from Butduu, and Goowt married Mab. Aleon's first name, by elimination, is Iaman. By clue 2, then, Mab's last name is Uwood, and Yano is from Weir. The female from Conphes is not Watta (clue 4), so she is Pherst and Watta was the fifth to be married. Hy is not from Lykta (clue 1), so Mab Uwood is, and Hy is from Telme. By clues 3 and 7, the male from Chother can only be Perfec, and Pherst's last name is Imuss. Hy's last name is not Ophens (clue 10); it is Ther, and Nouh's is Ophens. Ru's last name is not 4ee (clue 9); it is Frendle, and Perfec's is 4ee. Kis Ntel is not from Sheram (clue 6); he is from Ovkorsnot, and Ru Frendle hails from Sheram. In sum, in the order wed:

FEMALES	PLANET	MALES	PLANET
Hy Ther	Telme	Ru Frendle	Sheram
Nouh Ophens	Butduu	Kis Ntel	Ovkorsnot
Mab Uwood	Lykta	Goowt Withme	Idluvit
Pherst Imuss	Conphes	Iaman Aleon	Som'i
Watta Yano	Weir	Prefec 4ee	Chother

56. EDUCATIONAL OPPORTUNITIES

The classes are a pottery class on Tuesday (clue 3), a photography class on Wednesday (clue 6), a drawing class on Friday (clue 8), and two of the girls' classes—which must be drama and woodworking—on Saturday (clue 5). Leonard's music lesson is on Monday (clue 7), so Jonathan's is on Saturday (clue 5). The flute lesson is on Tuesday (clue 3), the violin lesson on Thursday (clue 7). No child has a music lesson and class on the same day, so Jonathan's class is Friday or earlier. His class is after Martha's music lesson and the guitar lesson (clue 1). Martha's music lesson isn't on Monday, and the guitar lesson isn't on Thursday (clue 7). By clue 1, then, the Friday drawing class is Jonathan's, the guitar lesson is on Wednesday, and the Tuesday flute lesson is Martha's. By elimination, Leonard's Monday music lesson and Jonathan's Saturday music lesson are either clarinet or piano. The child who takes clarinet lessons attends the photography class on Wednesday (clue 6), and can only be Leonard; Jonathan, by elimination, takes piano lessons. Karen doesn't take violin lessons (clue 2), so she takes guitar lessons and Nanette violin lessons. By clue 4, Karen also attends the woodworking class. Since Martha takes her music lesson on Tuesday, the Tuesday pottery class is Nanette's, and Martha attends the Saturday drama class. In sum:

Mon.: Leonard, clarinet
Tues.: Martha, flute; Nanette, pottery
Wed.: Karen, guitar; Leonard, photography
Thurs.: Nanette, violin
Fri.: Jonathan, drawing
Sat.: Jonathan, piano; Karen, woodworking; Martha, drama

57. BOBBING FOR APPLES

The Strom child didn't get any apples (clue 8). The two dressed as the hobo and the black cat didn't get the same number of apples (clue 7), so by clue 1, Eric and the Wick child got different numbers of apples, and the five children are the Strom child and the four mentioned in clue 1: Eric, the Wick child, the one dressed as the black cat and the one dressed as a hobo. Either Eric or the Wick child got the most apples. The Strom child who got no apples (clue 8), didn't come as a ghost (clue 5), witch (clue 6), hobo, or black cat (clue 1); the Strom child came as a vampire. The Strom child isn't Jane (clue 3), Adam (clue 4), Eric (clue 1), or Susan (clue 9) and must be Kari. The one dressed as a witch didn't get the most apples (clue 6), nor did the cat or the hobo (clue 1), so the one dressed as a ghost did; that child got six apples (clue 5). Susan wasn't the ghost since she didn't get the most apples (clue 6), she wasn't the witch (also clue 6), or the hobo (clue 2); she was the one dressed as a black cat. Now we know that either Eric or the Wick child came as a ghost and got the most apples. If the Wick child got six, Adam couldn't be Wick (clue 4), and so must be the one dressed as a hobo and he would have gotten three. Eric (who would then be the one dressed as a witch) would have gotten either four or two, and Susan (the cat) either two or one. Then the only place left is for the Blake girl to be Susan, the cat, who got one or two apples. But the Blake girl got more apples than Adam (clue 4), therefore, it must have been Eric who came as a ghost and got six apples and the Wick child was the witch. Now, by clue 1, Susan, dressed as a black cat, got three; the Wick child got four or two; and the child dressed as a hobo got two or one. By clue 6, the Wick child, dressed as the witch, must have gotten four apples, while the one who came as the hobo got two. Now, by clue 4, Adam can only be the child who came as the hobo, while the Blake girl is Susan. The Anderson child isn't Adam (clue 9) and must be Eric. By elimination, the Wick child is Jane, and Adam's last name is Vail. In sum:

Eric Anderson, ghost, 6
Jane Wick, witch, 4
Susan Blake, black cat, 3
Adam Vail, hobo, 2
Kari Strom, vampire, 0

58. FORGETFUL FOOTES

One woman turned them back third and the other at Flag Street (clue 6). By clue 3, the one who turned back at Flag was the first turnaround. By clue 7, they returned for the fiddle just before the turnaround at Frond and just after the one requested by Flo. Since Flo's turnaround was either first or third, then the Frond Street turnaround was either third or fifth. The return from Frond wasn't fifth (clue 4), so the order was Flo's turnaround first, the return for the fiddle second, and the Frond Street turnaround third. Francine, then, ordered the third return on Frond and Flo had the first turnaround on Flag. By clue 1, the Flag Street turnaround was followed by the second turnaround where a man fetched the fiddle, followed by the street where Fred said to turn around. Flag Street (clue 3), was preceded by Fork Street, which was preceded by the street where they turned around to fetch the flashlight. By clue 5, since Francine was third, Frank's request came fourth. It was one block after Francine's Frond Street turnaround, so he must have returned via Fork and Francine made her request to find the flashlight. Fred did not request the second return (clue 1), so Felix was the man who went back for the fiddle and Fred requested the final turnaround. By clue 2, Felix turned around at Fig and Frank returned for film. The fifth street they turned around on, by elimination, was Field, the street after Fig. By clue 4, the last turnaround wasn't for the fan; it was for the food and the first trip back was for the fan. In sum, in order of streets from the flat to the freeway:

Frond St., Francine, flashlight, third return
Fork St., Frank, film, fourth return
Flag St., Flo, fan, first return
Fig St., Felix, fiddle, second return
Field St., Fred, food, fifth return

59. WHO SAW THE GOAT?

Each child saw at least 1 cow, 1 horse, and 1 sheep, but no more than 9 of each, plus one other type of animal (introduction). Jeremy saw 1 horse (clue 4). Since no two saw the same number of any one type of animal (clue 1), Julie saw 2 or 3 horses and 6 or 9 sheep (clue 3). She didn't see the most sheep (clue 7), so Julie saw 6 sheep and 2 horses. Since Jennifer saw twice as many horses as sheep (clue 2), she saw no more than 4 sheep, so neither Jennifer, who saw fewer sheep than Julie, nor Julie was the child who saw the most sheep and the deer (clue 7), nor was Jeremy, who saw 3 llamas (clue 4); Jonathan was. The child who saw the most horses wasn't Jonathan or Julie (clue 7) or Jeremy (clue 4); Jennifer saw the most horses. Twelve horses and 12 cows were seen (clue 8); the only possible combinations to total 12 are: 1, 2, 3, 6, and 1, 2, 4, 5. Of the 12 horses, Jeremy saw 1, Julie 2, and Jennifer saw an even number (clue 2) that was larger than the number Jonathan saw (clue 7), so Jennifer saw 6 horses and 3 sheep (clue 2), and Jonathan saw 3 horses. Of the 12 cows, Jennifer saw the fewest (clue 7), so she saw 1 cow plus 1 jack rabbit or goat (clue 6) for a grand total of 11. Julie saw the most cows (clue 7)—either 5 or 6 (clue 8), but only one more than Jeremy (clue 5), so Julie saw 5 cows, Jeremy 4, and Jonathan saw 2. Julie, then, saw 13 animals plus the jack rabbit or goat to total 14, so she saw the jack rabbit and had the highest total (clue 6), Jennifer, then, saw the goat, and the boys saw 12 and 13 (clue 1). Jonathan's 2 cows, 3 horses, and 1 deer, plus at least 7 sheep (clue 7) total at least 13, so Jonathan saw exactly 7 sheep and his total is 13, while Jeremy saw 4 sheep to total 12. In summary:

Jennifer: 1 cow, 6 horses, 3 sheep, 1 goat; total 11
Jeremy: 4 cows, 1 horse, 4 sheep, 3 llamas; total 12
Jonathan: 2 cows, 3 horses, 7 sheep, 1 deer; total 13
Julie: 5 cows, 2 horses, 6 sheep, 1 jack rabbit; total 14

Mr. Nelson had neither Mr. Kerwin (clue 3) nor Mr. Taylor (clue 4) for a substitute, but one of the women. Mr. Lynch also had a woman sub (clue 8). Also, neither man was replaced by Ms. Windsor (clue 2). Mr. Lynch didn't have Ms. Clark (clue 8); Mr. Lynch had Ms. Block and Mr. Nelson had Ms. Clark. Ms. Morgan didn't have Mr. Taylor or Ms. Windsor as a sub (clue 6), she had Mr. Kerwin. The one out five days was not Mr. Nelson, Ms. Morgan (clue 3), Ms. Davis (clue 7), or Mr. Lynch (clue 8); Ms. Foley was. Ms. Windsor and Mr. Taylor were the only subs on Wednesday (clue 6), but by clue 2, Ms. Windsor didn't work on Monday or Tuesday. Mr. Taylor, then, must have been Ms. Foley's sub and Ms. Windsor replaced Ms. Davis, the math teacher (clue 2). Neither Ms. Davis nor Ms. Morgan was the teacher absent one day (that was a man, clue 5) or two days (clue 6); they were out three and four days. Ms. Davis was present on Monday and Tuesday (clue 2), since her sub wasn't in, so, Ms. Davis, then, was the teacher out on Wednesday, Thursday, and Friday and Ms. Morgan was out four days. By clue 6, she was present on Wednesday, since her sub wasn't in, so she was out the rest of the week. Mr. Lynch's sub worked fewer days than Mr. Nelson's sub (clue 8), so Mr. Lynch was out one day and Mr. Nelson two. Both Lynch and Nelson were out Tuesday (clue 3). Since only Ms. Morgan and Mr. Lynch weren't out on consecutive days (clue 5) and Mr. Nelson's sub wasn't in on Wednesday (clue 6), Mr. Nelson was out on Monday and Tuesday. Since Ms. Davis' sub was not in on Tuesday, she is not mentioned in clue 3. By substituting in clue 3, the Latin teacher can only be Mr. Lynch. The English teacher absent one more day than Ms. Davis (clue 7) must be Ms. Morgan. The history teacher, by clue 1, couldn't be Ms. Foley, who was out all week; so he's Mr. Nelson and Ms. Foley, by elimination, teaches art. In sum:

> Mr. Lynch: Latin, Tues. sub Ms. Block
> Mr. Nelson: history, Mon., Tues. sub Ms. Clark
> Ms. Davis: math, Wed., Thurs., Fri. sub Ms. Windsor
> Ms. Morgan: English, Mon., Tues., Thurs., Fri. sub Mr. Kerwin
> Ms. Foley: art, Mon.–Fri. sub Mr. Taylor

61. PLAYGROUND DUTIES

In the two week (10 day) period, with 5 duties per day, there were 50 duties. All but the third-grade teacher had three duty-free days in that 10 days, and no teacher had more than one duty daily (clue 1). Thus six teachers each had 7 duties and the third-grade teacher had 8. To keep any teacher from having more than one duty per day, changes on the first draft were made on the first-week recess and lunch duty lists and on the second week recess and noon duty lists (clue 2). Only K-3 teachers had recess and lunch duties (introduction); they are Baker, Evans, Anderson (clue 2), and Ferguson (clue 3). Only the kindergarten teacher pulled two recess duties the first week (Mon. & Fri.), so Ferguson teaches kindergarten (clue 3). Either Anderson and/or Evans must have been listed on the first draft for both bus and recess duties the second week (clue 2). Since Evans had no bus duty the second week, Anderson was listed for both bus duty and recess duty on Wednesday when the third-grade teacher was listed, so Anderson teaches third grade. Baker and Evans, then, teach first and second grades, and first-draft conflicts required they be switched on both recess and lunch duty lists the first week (clue 2). If Baker taught second grade and Evans taught first grade, there would be no conflicts on the schedule. But clue 2 says there were conflicts, so Baker must teach first grade and Evans second. To eliminate the conflict, Baker and Evans switched Tuesday and Wednesday recess duty. For lunch duties, Evans was first assigned to Monday and Friday; but only the Friday duty conflicted, so Baker and Evans switched Thursday and Friday lunch duty (clue 2). The second week, Anderson and Evans switched Tuesday and Wednesday recess duty. Among the upper-grade teachers, the only conflict was with Carter and/or Dewey the second week (clue 2). Dewey had no bus duty the second week, so the conflict was on Friday when Carter pulled bus duty and the fourth-grade teacher was listed for noon duty; thus Carter teaches fourth grade. Grayson pulled Thursday afternoon duty the second week (clue 4), at the time the sixth-grade teacher had noon duty, so Grayson doesn't teach sixth grade. Grayson teaches fifth and Dewey, by elimination, teaches sixth. Since Carter pulled Monday afternoon duty the second week (clue 4), Carter's Friday noon conflict was switched with Dewey's Thursday noon duty the second week (clue 2). From clue 1, Anderson, grade 3, had four duties each week; the other teachers each had four duties one week and three the other week. Anderson had three first-week duties (Mon.

bus, Tues. lunch, Thurs. recess) and four second-week duties (Mon. & Fri. lunch, Tues. recess, Wed. bus), so Anderson's eighth duty had to be the first week, but not on Friday (clue 5); Anderson had afternoon duty on Wednesday the first week (clue 1). Baker, grade 1, had three first-week duties (Tues. bus, Wed. recess, Fri. lunch) and four second-week duties (Mon. & Fri. recess, Wed. lunch, Thurs. bus), so had no afternoon duties. The same is true for Carter, grade 4, who had three first-week duties (Mon. & Thurs. noon, Wed. bus) and four second-week duties (Mon. afternoon, Tues. & Thurs. noon, Fri. bus). Evans, grade 2, had four first-week duties (Mon. & Thurs. lunch, Tues. recess, Fri. bus); Ferguson, the kindergarten teacher, had three first-week duties (Mon. & Fri. recess, Wed. lunch) and two duty-free days (clue 3), so Dewey (gr. 6) and Grayson (gr. 5) had to complete the first-week afternoon duty list. Grayson had Tuesday and Friday noon duties, so Dewey had afternoon duties those days to make four first-week duties (including Wed. noon & Thurs. bus). Grayson, then, had the Monday and Thursday afternoon duties to make four first-week duties plus three second-week duties (Tues. bus, Wed. noon, Thurs. afternoon) to total seven. Three second-week afternoon duty slots remain to be filled by Dewey (gr. 6), Evans (gr. 2), and Ferguson (K). Dewey had Friday noon duty and no Tuesday duties (clue 5), so was assigned Wednesday afternoon duty. Ferguson had lunch duty Tuesday, so Evans had Tuesday afternoon duty, and Ferguson, by elimination, had Friday afternoon duty. In summary:

1st wk.	Bus	Recess	Lunch	Noon	Afternoon
Mon.	Anderson, 3rd	Ferguson	Evans	Carter	Grayson
Tues.	Baker, 1st	Evans	Anderson	Grayson	Dewey
Wed.	Carter, 4th	Baker	Ferguson	Dewey	Anderson
Thurs.	Dewey, 6th	Anderson	Evans	Carter	Grayson
Fri.	Evans, 2nd	Ferguson	Baker	Grayson	Dewey
2nd wk.					
Mon.	Ferguson, Kdgn.	Baker	Anderson	Dewey	Carter
Tues.	Grayson, 5th	Anderson	Ferguson	Carter	Evans
Wed.	Anderson	Evans	Baker	Grayson	Dewey
Thurs.	Baker	Ferguson	Evans	Carter	Grayson
Fri.	Carter	Baker	Anderson	Dewey	Ferguson

62. TREASURE ISLAND

Bones' chest, which contained sand (clue 2), was at site C or E (clue 6), while the chest of coins was at site A or C (also clue 6). By the diagram, if Bones' chest was at site C, it would have been equidistant from every other chest. By clue 7, then, Bones' chest was not at site C; it was at site E. If the chest of coins were at site C, it also would have been equidistant from every other chest; by clue 1, then, it was not at site C, so it was at site A. The *Golden Goat* was at site B or C (clue 4) and the gems were at site C or D (also clue 4). By the same reasoning as above and clue 7, the *Golden Goat* wasn't at site C, so it was at site B. Since Bones' chest of sand was at site E and the coins at site A, by clue 4, the only site nearer the *Golden Goat* is for the arms to be at site C, leaving the gems to be at site D. By elimination, the ingots were at site B. Redbeard's chest was at site D (clue 5) and so contained the gems. If Pegleg's chest was at site C, the chest of coins and the *Golden Gull*'s chest would have been equally distant from it, contradicting clue 1. Pegleg's chest did not contain coins (clue 1), so it was not at site A; Pegleg's chest was, then, the *Golden Goat*'s at site B. Therefore, according to clue 1, the *Golden Gull*'s chest was at site D and so was Redbeard's ship, and the *Silver Skull*'s chest was the one containing the arms at site C. By clue 3, the *Golden Gleam*'s chest was at site A or E, as was Blackbeard's. Since Bones' chest was at site E, his ship must be the *Golden Gleam* and Blackbeard's chest was at site A. By elimination, the chest of arms at site C was Flint's and Blackbeard's ship was the *Silver Snail*. In sum:

site A: Blackbeard, *Silver Snail,* coins
site B: Pegleg, *Golden Goat,* ingots
site C: Flint, *Silver Skull,* arms
site D: Redbeard, *Golden Gull,* gems
site E: Bones, *Golden Gleam,* sand

63. PETE'S PAINT SHOP

All ten original car colors, nine new paint colors, and six interior colors are mentioned in the introduction or the clues. Five of the cars were painted in the morning—one each Monday through Friday; the other five were painted in the afternoon (introduction). David's car was painted yellow during a morning (clue 8), but not Monday morning (clue 14), Tuesday when Pete saw no men, nor Wednesday or Thursday when Pete saw no yellow cars (clue 10), it was painted Friday morning. Brad's car was painted no later than Wednesday (clue 3). Phil's tan car was painted Thursday morning or before (clue 6), and Ken's was painted before Phil's (clue 14) on Wednesday or before. Since Pete saw two men on Thursday (clue 10), only Phil and Nick can fill the Thursday positions, with Phil in the morning (clue 6). Mr. Peterson's car was painted Wednesday or before (clue 1), so he is not Phil, Nick, or David, nor (from clue 14) is he Ken. Brad is Peterson. Since there are five women named and three of them had their cars painted in the afternoon (clue 19), the two women whose cars were painted in the morning were Linda (clue 1) and Janet (clue 4). Linda's car was not painted Monday or Friday (clue 1), or Thursday (clue 10); since Brad Peterson's was not painted Tuesday (clue 10), Linda's was not painted Wednesday (clue 1). Thus, from clue 1, Linda's car was painted Tuesday morning, Brad Peterson's Monday, and the blue car was painted white on Wednesday. Since Nick's car was painted Thursday, Ms. Quincy's car was painted Friday afternoon, and Baker's on Wednesday (clue 2). From clue 3, Miller's was the other car to be painted on Wednesday. Brad's and Miller's cars were painted the same color, but not white or red (clue 3), yellow (clue 10), brown, blue, or green (clue 16). Since two were painted white and two yellow, the only two colors left are black and the unknown color, so both are black. Baker's, then, is the blue car that was painted white. Since Phil's car was painted Thursday, Lowell's was painted Friday (clue 6), so David is Lowell. Phil's car doesn't have a black interior (clue 6), so he is not Curry or Dawson (clue 13). Neither Phil nor Nick is Nixon, whose car was painted the same day a yellow car was (clues 7, 10), or Warner, whose car was painted yellow (clues 8, 10). Phil is Owens, and Nick is Curry or Dawson, so his car has a black interior (clue 13). Michelle is not Nixon (clue 7), Curry or Dawson (clue 13), Quincy (clue 17), or Warner, whose car was painted in the morning (clue 8). Michelle is Baker or Miller and thus had her car painted Wednesday afternoon (clue 19). Michelle's is not the white car (clue 7), so hers is the blue car that was painted white and she is Baker; Miller's then is the white car that was painted black on Wednesday morning. The red car that was painted white and the black car that was painted red were painted two days apart (clue 11); since neither was painted Wednesday, the red car was painted white on Tuesday and the black car was painted red on Thursday and must belong to Nick. Miller's white car and Brad's car were the only two painted black, so Brad's was originally gray, and thus was painted Monday afternoon (clue 12). Nixon's car and the yellow car were painted the same day (clue 7); the only day left for them is Tuesday. Linda is not Nixon (clue 18), so hers is the yellow car, and Nixon's red car was painted white Tuesday afternoon. The only afternoon left for the gray car that was painted blue (clue 12) is Friday. Donna's car and the green car were painted the same day (clue 5)—and again only Friday is left open, so Donna is Quincy and David's is the green car that was painted yellow. Donna Quincy's gray car was painted blue and has gray interior (clue 5). Christine, then, is the Nixon woman whose car was painted Tuesday afternoon (clues 4, 10). Linda's yellow car was not painted yellow (clue 16), so she is not Warner (clue 8) or Dawson (clue 18). She is Curry and Nick is Dawson. By elimination, Warner's car was painted yellow Monday morning (clue 8). Also by elimination, it was originally brown and has brown interior (clue 9), so Warner is not Ken, whose car has black interior (clue 13); Warner is Janet and Ken is Miller. Phil Owens, then, has a car with brown interior (clue 4) which was painted brown (clue 9). Linda's car, by elimination, was painted green. The two cars whose interior is the same as the new paint color (clue 15) belong to Ken Miller (black) and Phil Owens (brown). The only two whose original and/or new colors don't match the interior colors are Linda's and David's (clue 15). The other six interiors match the original paint color (also clue 15). We have already ascertained that Janet, Nick, and Donna are three of them, so Brad's interior is gray, Christine's red, and Michelle's blue. By elimination, David's car has the white interior. In summary:

Monday morn.	— Janet Warner, brown to yellow, brown int.
aft.	— Brad Peterson, gray to black, gray int.
Tuesday morn.	— Linda Curry, yellow to green, black int.
aft.	— Christine Nixon, red to white, red int.

Wednesday morn. — Ken Miller, white to black, black int.
    aft. — Michelle Baker, blue to white, blue int.
Thursday morn. — Phil Owens, tan to brown, brown int.
    aft. — Nick Dawson, black to red, black int.
Friday morn. — David Lowell, green to yellow, white int.
    aft. — Donna Quincy, gray to blue, gray int.

64. PARADES AND PICNICS

By clue 1, three families went to the picnic: the Powells, including their daughters Kathie and Candie; Kent and Eve and their children; and Mark and his wife and children. By clue 2, the other three families, who attended the parade only, are the Pierces, Eric and his wife and children, and Joey and his parents and sister. Clue 3 describes four distinct families who went to the parade, so at least one of these families must have attended the picnic as well. Among the four families, there are six daughters; if the Powells were not one of these families, there would be a total of eight girls among the children—but there are only seven, so the Powells are among the four who saw the parade, and Mr. Powell must be either John or Gary. Kay, who isn't Mrs. Peck and didn't attend the parade (clue 4), must be Mark's wife. Listing only the fathers among the families in clues 1 and 2, we have Mr. Pierce, Eric, Joey's father, Mr. Powell, Kent, and Mark; by clue 3, both John and Gary have only daughters, so Mr. Pierce, like Mr. Powell, must be either John or Gary. Joey's father must be Bill. Joey's sister isn't Tracie (clue 2), so Ann, her son and daughter Tracie who attended the parade (clue 3), belong to Eric's family. Their son must be Bobby who attended the parade only (clue 5). Cindie Peck didn't attend the parade (clue 5); she is not Kay's child (clue 4), so she must be Kent and Eve's daughter. Vickie Pyle, who went to the parade (clue 3), must be Joey's sister. May, who has a daughter named Abbie (clue 6), must be Mrs. Pierce and by elimination, her second daughter is Carrie. Gary isn't married to May (clue 3) so John is, and Gary is Mr. Powell. Ava attended the parade only (clue 5) and must be Bill's wife. By elimination, Dot is Mrs. Powell. All seven girls are now accounted for; Mark and Kay must have two sons. Robby is not one of them (clue 1) so Danny and Andy are their children and Robby is the Peck's second child. The Paynes have a daughter (clue 6); they must be Eric and Ann and by elimination, Mark and Kay are the Plums. In sum:

> Payne: Eric, Ann, Tracie, Bobby, parade only
> Peck: Kent, Eve, Cindie, Robby, picnic only
> Pierce: John, May, Abbie, Carrie, parade only
> Plum: Mark, Kay, Andy, Danny, picnic only
> Powell: Gary, Dot, Kathie, Candie, parade and picnic
> Pyle: Bill, Ava, Vickie, Joey, parade only

65. WEATHER REPORTS

No two cities had the same high, the same low, or the same difference between high and low (clue 1). Since Jack and Rebecca Hill's city had a lower low than two other cities (clue 5), it was one of the ones mentioned in clue 3, so Jack and Rebecca Hill live in Indianapolis. Mrs. River's city had a high-low difference that was 4° less than Wichita's (clue 4), 1° less than Mrs. Forest's city, and 2° less than Philadelphia's (clue 6). That is four of the five cities; the fifth is Indianapolis where the Hills live. The city with temperatures of 62°–42° and a 20° difference was not Indianapolis (clue 9) or Mrs. Forest's city (clue 11). No temperature was below 33° (clue 10); if Mrs. River's city had had a low of 42°, Wichita's would have been 28° (clue 4); if Philadelphia had had a low of 42°, the Hills would have had a low of 32° (clue 5). By elimination, Wichita had 62°–42°. Mrs. River's city, then, had 72°–56° (clue 4). Since Wichita had the 20° difference, Philadelphia's was 18°, Mrs. Forest's was 17°, and Mrs. River's was 16° (clues 4, 6). The city with the lowest high and lowest low (clue 8) had a difference of 21°, i.e., 49 minus 28 (clue 7), so, by elimination, was Indianapolis. The Hills, then, had a low of between 33 (clue 10) and 39—i.e. the highest possible of 88 (clue 10) minus 49 (clue 7) = 39; Felicia's low was 5° higher than the Hills' (clue 5), so was between 38° and 44°. Felicia, then, is not River with a low of 56° or Forest (clue 11), nor is she from Philadelphia (clue 5). Felicia is from Wichita, whose low was 42°, so the Hills had a low of 37°, a difference of 21°, so the high was 58° (clues 5, 7), and Philadelphia had a low of 47°, a difference of 18°, so the high was 65° (clues 5, 6). The low in Mrs. Forest's city was above 60° (clue 11), so it was the highest low and had the highest high (clue 8); the high was 49° above the lowest low of 37° (clue 7), so was 86°, the difference was 17°, so the low was 69°. Since Philadelphia had the third highest high, Mrs. Stone lives there (clue 2), and, by elimination, Mrs. Valley lives in Wichita and is Felicia. By clue 3, Mrs. Stone in Philadelphia (which had the third-lowest low) must be Brenda. Diane is not Forest (clue 11), so Catherine is, and Diane is River. Catherine is not from San Francisco (clue 2), so she is from Tampa, and Diane is from San Francisco. In summary:

Indianapolis, 58°–37°, Rebecca Hill
Philadelphia, 65°–47°, Brenda Stone
San Francisco, 72°–56°, Diane River
Tampa, 86°–69°, Catherine Forest
Wichita, 62°–42°, Felicia Valley

66. RANKING THE PLAYERS

Ned and Eames are, in some order, best and worst batter (clue 11). By clue 4, Ned is not the best batter; he is the worst, or #7, while Eames is #1 in batting. Sam is not Eames (clue 7), so he is a third player. Dunn is not Sam or Ned (clue 7), so he is a fourth player. Dunn is either first or seventh in fielding (clue 13). He is not seventh (clue 7), so he is first. The worst fielder is fourth in batting (clue 13), so he is not Ned, Eames, or Dunn. He is not Sam (clue 7); he is a fifth player. In batting, Sam placed lower than Dunn, who placed lower than the average batter (clue 7); since Ned placed seventh in batting, Dunn placed fifth and Sam sixth. By clue 14, Fry is Ned, who placed seventh in batting. In fielding, Dunn placed first and the average batter placed seventh. By clue 4, then, Ned Fry placed, at best, third in fielding. He did not place fourth (clue 8), or sixth (clue 1). By clue 14, then, Sam did not place third or fifth. He did not place sixth (clue 5), so Sam placed second or fourth. By clue 10, Mel is not Eames, the #1 batter, or Dunn, the #1 fielder, or the worst fielder who batted #4; Mel is #2 or #3 on both lists. If Sam is #2 fielder, Ned is #3 (clue 14), leaving no place for Mel, so Ned is #3, Sam is #4, and Mel is #2 and the sixth player. The seventh player, by elimination, placed third in batting and fifth or sixth in fielding. By clue 4, Mel is right fielder Archer. Brown is not Sam or the fourth batter (clue 5); he is the seventh player, who placed third in batting. Clark placed higher in batting than Dunn (clue 2); he is not, then, Sam, so he is the player who placed fourth in batting and seventh in fielding. By elimination, Sam is Grant. Clark is not Jim (clue 1), Hal (clue 2), or Ken (clue 14); he is Phil. By clue 14, Ken is Brown, and he ranked sixth in fielding. Jim is not Eames (clue 1), so he is Dunn and, by elimination, Hal is Eames, who ranked fifth in fielding. Hal Eames and Phil Clark are both either outfielders or basemen, as are Jim Dunn and Ned Fry (clue 6). By clue 9, since right fielder Mel Archer placed second on the fielding list, Jim Dunn is not an outfielder; he and Fry are basemen, while Eames and Clark are outfielders. Eames is not the center fielder (clue 3), so he is the left fielder, while Clark, by elimination, is the center fielder. Since Sam Grant placed sixth in batting and baseman Jim Dunn placed fifth, Sam is not a baseman (clue 12). Since Sam placed fourth in fielding and left fielder Hal Eames placed fifth, Sam is not a fielder (clue 9); Sam Grant is the shortstop. Since center fielder Phil Clark placed fourth in batting and seventh in fielding, Ken Brown is the second baseman (clue 3). Ned Fry does not play first base (clue 8), so he plays third base and, by elimination, Jim Dunn plays first. In summary, with batting rank listed first:

First baseman Jim Dunn, fifth; first
Second baseman Ken Brown, third, sixth
Third baseman Ned Fry; seventh, third
Shortstop Sam Grant, sixth, fourth
Left fielder Hal Eames, first, fifth
Center fielder Phil Clark, fourth, seventh
Right fielder Mel Archer, second, second

67. THE HOUNDING AT BASKER HILLS

Gwendolyn found the third item (clue 6). The character who found the first item was not Arden (clue 2), Duncan (clue 3), or Rayna (clue 5), so he or she was either Leatrice or

Micah. Leatrice's item was found just before a second woman's (clue 7) so, if Leatrice had found her item first, Rayna would have found hers second, and Gwendolyn would have been third. Rayna, however, found her item after two men had found theirs (clue 5); Leatrice was not first, so Micah was. The person who found the second item was not Arden (clue 2) or Rayna (clue 5); he or she was either Duncan or Leatrice. The person who found the last item was not Arden (clue 4) or Leatrice (clue 7); he or she was either Duncan or Rayna. If Duncan had found the second item, then Rayna would have found the sixth. Leatrice, then, would have found the fifth (clue 7); Arden, by elimination, would have found the fourth. Micah would have found the sword (clue 3). Rayna found her item in either the dining hall or the foyer (clue 5); if she had found the sixth item, she would not have found it in the dining hall (clue 4), so she would have found it in the foyer. The item she would have found would not be the potion (clue 1), the cloak, the password (clue 2), or the staff (clue 4); she would have found the magic stone. Arden, who would have found the fourth item, would not have found the cloak, the password (clue 2), or the staff (clue 9); he would have found the magic potion. Leatrice, who would have found the fifth item, would not have found the cloak or the password (clue 2); she would have found the staff. However, since the cloak was found by neither Gwendolyn (clue 6) nor Duncan (clue 3), this would leave no one to find the cloak; Duncan was not the one who found the second item; Leatrice was. Arden did not find the sixth item (clue 4), so either Rayna or Duncan did. Since two men found items before Rayna (clue 5), if Duncan had found the sixth item, Arden would have found the fourth item, and Rayna the fifth. Arden and Micah, then, would have found, in some order, the password and the item in the tower (clue 5). Arden did not find the password (clue 2), so Micah would have found it, and Arden would have found the item in the tower. Rayna would have found the sword (clue 3). By clue 10, then, Duncan would have found the staff; this, however, would contradict clue 4, so Duncan did not find the sixth item. Rayna, then, found the sixth item. She found that item in either the dining hall or the foyer (clue 5); she did not find it in the dining hall (clue 4), so she found it in the foyer. That item was not the potion (clue 1), the cloak, the password (clue 2), the sword (clue 3), or the staff (clue 4); it was the magic stone. Duncan and Arden found, in some order, the fourth and fifth items. If Arden had found the fourth item and Duncan the fifth, Arden would have found the sword (clue 3), and the password would have been found by Leatrice, Gwendolyn, or Micah (clue 2); since it was found by a man (clue 5), it would have been found by Micah. The item in the tower would have been found by either Duncan or Arden (clue 5). Since the stone was found sixth, the item in the tower was not found fifth (clue 10); it would, then, have been found fourth by Arden. This, however, contradicts clue 8, so Arden did not find the fourth item; Duncan found the fourth item and, by elimination, Arden found the fifth. Gwendolyn, then, found the sword (clue 3). The dining hall was still being searched for its item after two men had found the password and the tower item (clue 5); Arden, then, found the dining hall item fifth, while Micah and Duncan found, in some order, the password and the tower item. Gwendolyn did not find the sword in the Presence Chamber (clue 6) or the kitchen (clue 7); she found it in the dungeon. Leatrice did not find her item in the kitchen (clue 7); she found it in the Presence Chamber. Duncan, who found his item fourth, did not find it in the kitchen (clue 9); he found it in the tower and, by elimination, Micah found his item in the kitchen. By clue 5, either Micah or Arden found the password. Arden did not find it (clue 2), so Micah found it. The cloak was not found by Arden (clue 2) or Duncan (clue 3); it was found by Leatrice. Duncan did not find the staff (clue 9), so Arden found that and, by elimination, Duncan found the magic potion. In summary, in the order the objects were found:

> Micah, password, kitchen
> Leatrice, cloak, Presence Chamber
> Gwendolyn, sword, dungeon
> Duncan, potion, tower
> Arden, staff, dining hall
> Rayna, stone, foyer

68. RIVERVIEW APARTMENTS

There is one single woman; she is Miss Hart (clue 6). Faith's family lives next-door (clue 6), and there is a vacant apartment also next-door (clue 10); Miss Hart lives in a center (B) apartment. The Carters live in a center (B) apartment, on either the third or fourth floor, and Oscar lives two floors directly below the Carters (clue 5). Both vacant apartments are on the second floor and above (clue 2). There is one single man; on his floor, there is a vacant apartment (clue 10) and one family with a child (clue 1). His last name, then, is not Engel (introduction), Downing (clue 3), Carter (clue 5), Gordon (clue 7), Jackson (clue 8), or Brown (clue 11); he is Wright, Alden, or Farrell. If he were Wright, there would be a vacant apartment directly below his, and Roger's apartment would be directly below that (clue 2). The Carters, then, would live in 4B, Mark/June in 4A and 4C, Oscar in 2B, Wright on the third floor in either A or C, the first vacant apartment would be next-door to Wright, the second vacant apartment would be on the second floor directly below Wright, and Roger would be on the first floor, directly below the vacant apartment. This, however, would leave no place for Miss Hart; Wright is not the single man. If the single man were Farrell, his floor would include Zack's apartment (clue 7) and a vacant apartment (clue 10). If the vacant apartment were apartment B, by the above reasoning, the Carters would live directly above it and Oscar directly below, and Miss Hart would live directly above the Carters. This, however, would leave no place for Roger or the Wright apartment (clue 2); if the single man is Farrell, the vacant apartment on his floor is not Apartment B. Farrell, then, would live in Apartment B (clue 10). Since Oscar is not Farrell (clue 5), by the above reasoning, the Carters would live directly above Farrell, Oscar would live directly below Farrell, and Miss Hart would live directly above the Carters. Farrell, then, would live in 2B, the Carters in 3B, Miss Hart in 4B, and Oscar in 1B. The vacant apartment on Farrell's floor, then, would be the apartment directly below the Wright apartment, and directly above Roger's (clue 2); the vacant apartment on Miss Hart's floor would be two floors directly above Sally's (clue 3). Sally and Zack, then, would be married. This, however, leaves no place for the Downing apartment on Sally's floor (clue 3); Farrell is not the single man. By elimination, the single man is Mr. Alden. Mr. Alden, Clara, and a vacant apartment are on one floor (clue 7). Miss Hart, a vacant apartment, and Faith's family live on another floor (clue 6). The Carters, June, and Mark are the tenants of a third floor (clue 5). The Carters live in apartment B on their floor (clue 5), so Bob is not Mr. Carter (clue 4). He is not June's husband (clue 5). Bob, the Gordons, and Paula are tenants on a fourth floor (clue 7). Sally and the Downings have no children (clue 3); since there is one child on every floor (clue 1), the third tenant on their floor is a family with one child. The Carters live on floor three or four (clue 5); Sally lives on floor one or two (clue 3). Sally and the Downings, then, are the same floor as Bob, Paula, and the Gordon apartment. Bob has a child (clue 11), so Sally is Mrs. Gordon, and Paula is Mrs. Downing. By clue 2, neither Mr. Alden nor Miss Hart live on the first floor; by clue 5, the Carters do not live on the first floor. By elimination, Sally Gordon, Bob, and Paula Downing live on the first floor. By clue 3, then, there is a vacant apartment on the third floor. The Carters, June, and Mark live on the fourth floor, and the Carters live in 4B (clue 5). Oscar lives in 2B (clue 5). Miss Hart, then, lives in 3B, with a vacant apartment on one side (clue 10) and Faith's family on the other (clue 6). Sally, then, does not live in 1B (clue 3); nor does Bob (clue 4), so Paula Downing lives in 1B. Mr. Alden, Clara, and a vacant apartment, by elimination, are on the second floor; by clue 10, Mr. Alden is Oscar. Clara does not live in apartment C (clue 4); she lives in apartment 2A, and 2C is vacant. The vacant apartment on the third floor is not directly above the vacant apartment on the second floor (clue 2); apartment 3A is vacant, and Faith's family lives in 3C. Sally Gordon, then, lives in 1A (clue 3), so Bob lives in 1C. By clue 2, then, Roger lives in 2A and is Clara's husband, the Wrights live in 4A, and Sam lives in 1C and is Bob's son. Zack and the Farrells live on the same floor (clue 7); they do not live on the first (clue 11) so, by elimination, they live on the fourth, and the Farrells live in 4C. Bob's wife is not Holly (clue 8), Nancy or Ann (clue 11), so she is either Doris or Laura. If she were Laura, Doris would live on the third floor and Holly on the fourth (clue 8). This, however, would leave no floor for Nancy and Ann (clue 11); Bob's wife is Doris. Their last name is not Jackson (Clue 8) or Brown (clue 11); it is Engel. Nancy has a son (clue 11), so she is not Faith's mother; she and Ann live on the fourth floor. By elimination, Laura and Holly are, in some order, Miss Hart and Faith's mother. Since Laura and Ed live on the same floor (clue 7), Laura is Miss Hart, and Holly and Ed are Faith's parents. They are not the Jacksons (clue 8), so they are the Browns and, by elimination, Clara and Roger are the Jacksons. Since Faith lives in 3C and Sam lives in 1C, by clue 1, then, Amanda and Vince live, in some order, on the second and fourth floors, in apartments A and B. Nancy, who lives on the fourth floor, has a son (clue 11); she is Vince's mother. Amanda, then, lives on the second floor (clue 1); she is Clara and Roger Jackson's daughter in 2A. Nancy and Vince, then, live in 4B (clue 1); they are the Carters. Ann does not live in apartment A (clue 4); she lives in apartment 4C and, by elimination, June lives in apartment 4A. Mark, then, is Ann Farrell's husband (clue 5). Zack does not live in 4B (clue 12); he lives in 4A and is June Wright's husband. By clue 9, George lives on the fourth floor; he lives in 4B and is Mr. Carter; Tom lives in 1B and is Mr. Downing (clue 9). By elimination, William lives in 1A and is Mr. Gordon. In summary:

164

1: A: Sally and William Gordon; B: Paula and Tom Downing;
   C: Doris, Bob, and Sam Engel
2: A: Clara, Roger, and Amanda Jackson; B: Oscar Alden; C: vacant
3: A: vacant; B: Laura Hart; C: Holly, Ed, and Faith Brown
4: A: June and Zack Wright; B: Nancy, George, and Vince Carter;
   C: Ann and Mark Farrell

69. LOGIC PROBLEM CONTEST

By the introduction, the first winning puzzle was published in January. By clue 10, two puzzles will be published in the spring, two in the fall, and Ivan's in the summer. One puzzle will be published in July (clue 6), and no puzzles will be issued in May or June (clue 1). By clue 5, a puzzle will be published the month preceding Ivan's, so there are two summer publications. Henry's, then, is the only puzzle published in a winter month (clue 10), so his was published first, in January. By clue 3 then, the puzzles were published in the following order: Henry's, the puzzle by a woman about dolls, the minister's puzzle, Fern's puzzle, the Peoria winner's puzzle, O'Reilly's puzzle, and the medium-level puzzle published in November. The puzzle about the dolls and the minister's puzzle were the two spring publications, Fern's and the one from Peoria were the two summer publications, and O'Reilly's and the medium puzzle published in November were the two fall publications. By clue 5, Fern is the woman from Little Rock, whose hard puzzle was published one month before Ivan's, and Ivan is from Peoria. Since one puzzle was published in July (clue 6), Fern had her puzzle published in July and Ivan's was published in August. The first fall publication was not in September (clue 1), so it was in October. There were no puzzles published in May or June (clue 1); by clue 7, the publication just before Fern's July publication was in April. Since February is a winter month, the second publication came out in March (clue 10). There were three women winners, Esther, Fern, and Jane. Fern's puzzle was published in July. Jane's was published in October or November (clue 10). The woman who wrote about dolls who was published in March, then, was Esther and, by elimination, the minister was a man. The minister's puzzle was published in April, so it was not the challenger (clue 7). Esther's was easier than the minister's (clue 12), so it was not the challenger. Since the two hards were by Ivan and Fern, the minister's puzzle was the second medium. Esther's, then, was an easy (clue 12). Henry's puzzle was not an easy (clue 12), so it was the challenger and, by elimination, O'Reilly's puzzle was the second easy. Esther's last name is not Quincy (clue 6), Minor (clue 7), Rush, Price (clue 9), or Lewis (clue 12); it is Nobel. Fern's last name is not Quincy, Price (clue 6), Minor (clue 7), or Rush (clue 9); it is Lewis. Ivan's last name is not Quincy (clue 6), Price, or Rush (clue 9); it is Minor. Mr. Quincy is not Ken or Henry (clue 6), so he is Greg. If Jane's puzzle had been published last, Greg Quincy's would have been published in April, and he would be the minister. He would not have been from Miami (clue 13), Denver (clue 4), Milwaukee (clue 6), or Seattle (clue 7), so he would have been from Boston. By elimination, Ken would have been O'Reilly, who wrote the puzzle about sweatshirts (clue 10), and Jane's last name would have been either Rush or Price. By clue 9, she would have written about either the track meet or the beauty contest. She would not, then, have been from Seattle (clue 9); Esther would have been the woman from Seattle and, by clue 4, Jane would have been the housewife from Denver. However, since she would have written about either the track meet or the beauty contest, this would contradict clue 4. Jane, then, is O'Reilly, and the puzzle published in November was about sweatshirts (clue 10). Neither Rush nor Price wrote about sweatshirts (clue 9), so Greg Quincy did. By elimination, Ken is the minister. Greg Quincy is not from Denver (clue 4), Milwaukee (clue 6), Seattle (clue 7), or Miami (clue 8); he is from Boston. Minister Ken is not from Miami (clue 13), Denver (clue 4), or Seattle (clue 7); he is from Milwaukee. His last name is not Price (clue 9), so he is Rush and, by elimination, Henry is Price. Henry Price does not live in Denver (clue 4) or Seattle (clue 7); he is from Miami. He is the sales manager (clue 13). Ken Rush wrote about either the track meet or the beauty contest (clue 9); he did not write about the track meet (clue 2), so he wrote about the beauty contest, and Henry Price wrote about the track meet (clue 9). Greg Quincy, who wrote the other medium, is not, then, the pediatrician (clue 12). Nor is he the schoolteacher (clue 2) or the librarian (clue 11); he is the track coach. Ivan Minor from Peoria is not the pediatrician (clue 7), or the librarian (clue 11); he is the teacher. By clue 12, either Fern is the pediatrician and Ivan wrote about the shopping mall, or Esther is the pediatrician and Jane wrote about the shopping mall. By clue 7, if Esther were the pediatrician, Jane would be the woman from Seattle. This, however, would leave no one to be the housewife from Denver (clue 4). Fern is the pediatrician and Ivan wrote about the shopping mall (clue 12). Fern did not write about teddy bears (clue 7), so Jane did and, by elimination, Fern wrote about the classroom seating chart. Jane is not the woman from Seattle (clue 7), so she is the housewife from Denver (clue 4) and, by elimination, Esther is the woman from Seattle, who is a librarian. In summary, in order of publication:

Jan.: Henry Price, sales manager from Miami, track meet, challenger
March: Esther Nobel, librarian from Seattle, dolls, easy
April: Ken Rush, minister from Milwaukee, beauty contest, medium
July: Fern Lewis, pediatrician from Little Rock, seating chart, hard
August: Ivan Minor, teacher from Peoria, mall, hard
Oct.: Jane O'Reilly, housewife from Denver, teddy bears, easy
Nov.: Greg Quincy, track coach from Boston, sweatshirts, medium

70. HANDCAR RALLY

By the introduction, there were six towns along the route; the first town was Summerset and the sixth Ocean City. The total distance of the rally was 120 miles, and each leg is at least ten miles long (clue 1). By clue 5, Jeff and his partner pumped either the first or second leg of the rally. If they pumped the first leg, then Lakeside would have been the second town and Riverfront the fourth (clue 5). By clue 9, then, Hilltop would have been the fifth town, Greg and his partner would have done the second leg of the rally, and the third leg would have been 12 miles long. By elimination, Mountainview would have been the third town. By clue 2, the distance from Summerset to Lakeside would have been 48 miles. By clue 4, Brad would not have pumped the third or fourth leg. If he had pumped the first leg as Jeff's partner, the fourth leg would have been 24 miles (clue 4). The total of the first, third, and fourth legs, then, would have been 84 miles, leaving 36 miles for the second and fifth legs. Since all five legs were at least ten miles long (clue 1), by clue 8, Chad did not pump the second or third leg, so he would have pumped either the fourth or fifth leg (clue 8). If he had pumped the fourth leg, the second leg would also have been 12 miles (clue 8), contradicting clue 1. If he had pumped the fifth leg, since the second and fifth legs would have totaled 36 miles, the second leg would have been 12 miles and the fifth 24, also contradicting clue 1. Brad, then, did not pump the first leg. If Brad had pumped the second leg, then Chad would have pumped twice as far as Brad (clue 8), who would have pumped twice as far as the fourth leg (clue 4). If Chad had pumped the first leg, Brad would have pumped 24 miles, and the fourth leg would have been 12 miles, contradicting clue 1. Chad then, would have pumped the fifth leg; the fifth leg would have been twice the second leg, which would have been twice the fourth leg. Since the fourth leg was at least ten miles, the total of these three legs would have been at least 70 miles, making the total of all five legs at least 130 miles, contradicting clue 1. So, if Jeff had pumped the first leg of the rally, Brad would have pumped the fifth. By clue 8, Chad would not have pumped the second or third leg. If he had pumped the fourth leg, the fourth leg would have been twice the distance of the second leg (clue 8) and the fifth leg would have been twice the distance of the fourth (clue 4), again leading to a total of at least 70 miles for these three legs. If Chad had pumped the fifth leg with Brad, the second and the fourth legs would have been the same length, contradicting clue 1. Chad, then, would have pumped the first leg with Jeff. The second leg, then, would have been 24 miles long (clue 8). Since the first three legs would have been 48, 24, and 12 miles long, the fourth and fifth legs would have totaled 36 miles in length (clue 1). By clue 4, then, the fifth leg would have been 24 miles and the fourth 12 miles, contradicting clue 1. So, if Jeff had pumped the first leg, there would have been no leg Chad and his partner could have pumped; Jeff did not pump the first leg of the rally. By clue 5, then, Jeff and his partner pumped the second leg, Lakeside was the third town on the route, and Riverfront was the fifth town. By clue 9, then, Greg and his partner pumped the first leg, the second leg was 12 miles long, and Hilltop was the fourth town on the route. By elimination, Mountainview was the second town on the route. Since Lakeside was 12 miles from Mountainview, Mountainview was 36 miles from Summerset (clue 2). By clue 8, Chad and his partner pumped 24 miles. If they had pumped the fourth leg, Brad and his partner would have pumped 48 miles (clue 4). Since the first leg was 36 miles and the second leg 12 miles, the total of just these four legs would have been 120 miles, contradicting clue 1. Chad and his partner, then, pumped the third or fifth leg of the rally. Brad and his partner did not pump the second or

166

fourth leg (clue 4). If they pumped either the third or fifth leg (whichever was not pumped by Chad and his partner), then, by the totals of the other legs, the fourth leg would have been 16 miles and the third or fifth 32 miles (clue 4). This, however, would contradict both clue 3 and clue 6; Brad pumped the first leg with Greg. The fourth leg, then, was 18 miles long (clue 4). To summarize, the first leg pumped by Greg and Brad was 36 miles; the second leg, pumped by Jeff and his partner, was 12 miles; the fourth leg was 18 miles; Chad's leg, which was either the third or the fifth, was 24 miles. The leg *not* pumped by Chad, then, was 30 miles (clue 1). There are four pairs of legs that are six miles apart: 36/30, 30/24, 24/18, and 18/12. One of these pairs, then, is the Danny/Ian pair mentioned in clue 6. Since Greg and Brad pumped the 36-mile leg, Danny did not pump that. Chad, who pumped the 24-mile leg, did not partner with Danny (clue 11). Jeff, who pumped the 12-mile leg, did not pump with Ian (clue 6). By elimination, then, Danny pumped the 30-mile leg, and Ian pumped the 24-mile leg with Chad. By clue 3, then, Herb and his partner pumped 18 miles and Ed and his partner pumped 12 miles—Herb pumped the fourth leg, while Ed pumped the second leg with his partner Jeff. The only men not yet paired are Adam and Fran. Adam did not pair with Herb (clue 10), so Adam was Danny's partner, while Herb's partner was Fran. Fran and Herb pumped the fourth leg, so Ian and Chad pumped after that (clue 7); they pumped the fifth leg so, by elimination, Danny and Adam pumped the third. In summary:

Summerset to Mountainview: Greg & Brad, 36 miles
Mountainview to Lakeside: Jeff & Ed, 12 miles
Lakeside to Hilltop: Danny & Adam, 30 miles
Hilltop to Riverfront: Herb & Fran, 18 miles
Riverfront to Ocean City: Chad & Ian, 24 miles

71. WEIRD PET RACES

By the introduction, the sixth-place pet scored 10 points, and the first-place pet scored, at most, 30 points. The three pets mentioned in clue 5, then, were either the first-, second-, and third-place pets, and the first-place pet scored, at most, 29, or they were the second-, third-, and fourth-place pets, and the first-place pet scored 30. The first-place pet was not the armadillo, the aardvark (clue 2), or the chameleon (clue 3). Beth's pet did not place first or second (clue 7), so the anteater did not place first (clue 1). Either the platypus or the iguana, then, placed first. If the platypus had placed first, then Buster would have placed second with seven fewer points (clue 6). The highest score was, at most, 30. If the platypus had scored 30 points, Buster would have scored 23 points. By clue 5, then, the third- and fourth-place pets would have scored 22 and 21 points respectively, contradicting clue 1. If the platypus had scored less than 30 points, then the third-place pet would have scored 21 points, the second-place pet would have scored 22 points, and the platypus would have scored 29 points (clues 5 and 6). The fifth-place pet would, then, have scored 11 points (clue 4). However, since the sixth-place pet scored 10 points, this would again contradict clue 1; the platypus did not place first. The iguana, by elimination, placed first. If the platypus had placed second, then Buster would have placed third and scored seven points less (clue 6). Since the iguana scored, at most, 30 points, and since the iguana scored more than 1 point more than the second-place pet (clue 1), the platypus would have scored, at most, 28 points. Buster, then, would have scored, at most, 21 points; this, however, would leave no place for the third pet who scored between 21 and 29 points (clue 5); the platypus did not place second. If the platypus had placed third, then Buster would have placed fourth and scored seven points less (clue 6). The fifth-place pet would have scored three points less than Buster (clue 4). By clue 5, then, the fourth-place pet could have not scored 21 points or more; the first-, second-, and third-place pets would have been those mentioned in clue 6: the first-place pet would have scored, at most, 29 points, the second-place pet would have scored, at most, 27 points, the third-place pet would have scored, at most, 26 points, the fourth-place pet would have scored, at most, 19 points, and the fifth-place pet would have scored, at most, 16 points. By clue 3, then, the chameleon would have placed sixth, and Trisha's pet would have placed fourth or fifth. By clues 6 and 7, then, Beth's pet would have finished fifth or sixth; by 1, the anteater would have placed fifth and Beth's pet would have been the chameleon. Trisha's pet, then, would have finished fourth. The anteater in fifth would not have been Emma's (clue 1), Chad's (clue 5), or Larry's (clue 9); it would have been Corey's. Its name would not have been Max (clue 3), Binky (clue 5), Star (clue 8), or Sparky (clue 9); it would have been Baby. Beth's pet would not have been named Max (clue 3), Binky (clue 5), Star (clue 8), or Sparky (clue 9); there would be no remaining name for

Beth's pet, so the platypus did not finish third. If the platypus had placed fifth, then Buster would have placed sixth and scored seven points less (clue 6); it would have scored 10 points, so the platypus would have scored 17 points. The third-place pet, then, would have scored 27 points (clue 4) so, by clue 5, the first-place pet would have scored 30 points, the second-place pet would have scored 28 points, and the fourth-place pet would have scored between 21 and 25 points. Beth's pet would have placed third, and the anteater second (clue 1). Trisha's pet, then, would be the anteater, and the chameleon would have finished fourth with a score of 24 (clue 3). Chad, then, would own the chameleon (clue 5). However, this leaves no room for Larry, whose pet didn't finish first (clue 9), fifth, or sixth (clue 6); the platypus did not finish fifth. By elimination, the platypus placed fourth and Buster placed fifth (clue 6). Since Buster scored ten points less than the third-place pet (clue 4) and seven points less than the fourth-place platypus (clue 6), the platypus scored three points less than the third-place pet. Trisha's pet did not place first, third, fourth, or sixth (clue 3). If it had placed second, the chameleon would have scored four points less, and would have placed third (clue 3). The fifth- and sixth-place pets, then, would have been the anteater and Beth's pet, whose final scores were one point apart (clue 1). Buster was not Beth's pet (clue 6), so Buster would have been the anteater, and Beth's pet would have placed sixth. Since Beth's pet would have scored 10 points, the anteater would have scored 11 points, the fourth-place pet would have scored 18 points, the third-place chameleon would have scored 21 points, Trisha's second-place pet would have scored 25 points, and the first-place pet would have scored between 27 and 29 points. Buster would not have been Emma's (clue 1), Chad's (clue 5), or Larry's (clue 6); it would have been Corey's. The platypus would not have been Chad's (clue 5) or Larry's (clue 6); it would have been Emma's. By clue 9, then, Larry's pet would have placed third, and would have been the chameleon, while Trisha's pet would have been named Sparky. By elimination, Chad's pet would have placed first. Its name would have been Max (clue 3). Larry's chameleon then, would have been named Binky (clue 5). Beth's pet was not named Star (clue 8), so Emma's platypus would have been, and Beth's pet would have been named Baby. By elimination, however, Beth's pet would be either the armadillo or the aardvark; this, then, would contradict clue 2. Trisha's pet then did not place second; it placed fifth and was Buster, and the chameleon placed sixth (clue 3). By clue 1, then, the anteater placed second, and Beth's pet placed third. Since the chameleon scored 10 points, Trisha's Buster scored 14 points (clue 3), the platypus scored 21 points (clue 6), Beth's pet scored 24 points (clue 4), the anteater scored 25 points (clue 1), and the iguana scored 30 points (clue 5). By clue 5, Beth's pet was named Binky, and Chad's pet was the fourth-place platypus. The owner of the iguana was not Corey (clue 8) or Larry (clue 9); she was Emma. By clue 9, then, Larry's pet was the anteater, and Emma's iguana was named Sparky. By elimination, Corey owned the chameleon. Neither Chad (clue 5) nor Corey (clue 8) owned Star; Larry did. By clue 3, Max was Chad's pet and, by elimination, Corey's pet was named Baby. Binky was not the armadillo (clue 5), so it was the aardvark and, by elimination, Trisha's Buster was the armadillo. In summary, in order of finish:

Emma's Sparky, iguana, 30
Larry's Star, anteater, 25
Beth's Binky, aardvark, 24
Chad's Max, platypus, 21
Trisha's Buster, armadillo, 14
Corey's Baby, chameleon, 10

72. SUMMERSET OUTLET MALL

By clue 1, the four corner shops are Ian's, Heidi's, the appliance shop, and the bath shop. Only two shops can be entered from Art's (clue 7); his is one of the corner shops, either the bath or the appliance shop. Don's shop and the bath shop do not connect (clue 6); Art's shop and Don's shop are connected (clue 7), so Art does not own the bath shop; he owns the appliance shop. Ian's and Art's Appliances are diagonally opposite each other, as are Heidi's and the bath shop (clue 1). If only one shop separated the bath shop and Art's (so that, going clockwise around the mall, the shops would be Ian's, Heidi's, Art's Appliances, and the bath shop), Don's shop would be between Heidi's and Art's, with an unknown shop between Heidi's and Don's, and the candle shop would be between Art's and the bath shop (clues 6 and 7). By clue 4, then, Jack's shop is one of the outer shops, and is between Ian's shop and the bath shop. Ian's shop is not the glassware shop (clue 9), so Fran's, Gail's, or the glassware shop is between Ian's and Jack's. That shop would not be the glassware shop (clue 9) or Gail's (clue 10); it would be Fran's. The bath shop, then, would be Gail's (clue 4), and the shop between Jack's and Don's would be the glassware shop. By clue 2, then, the artwork shop, which is one of the two center stores, would then be next to Fran's, and the glassware shop would be Bill's. Bill's and Jack's, then, would have a connecting door; this, however, contradicts clue 13. So two shops separate Art's and the bath shop; the order, going clockwise around the mall, is: Ian's, the bath shop, Art's Appliances, and Heidi's. Ellen's is store #7 (clue 8). The glassware shop is not Heidi's (clue 12). If Ian's was store #1, then Art's Appliances would be store #12, Heidi's would be store #10, and the bath shop would be store #3. Stores #9 and #11, then, would be, in some order, Don's and the candle shop (clue 7). By clue 4, then, Jack's would be store #6, so store #9 would be Don's glassware, store #11 would be the candle shop, while stores #3 and #5 would be, in some order, Fran's and Gail's. Since the artwork shop would not connect to Don's Glassware (clue 2), it would be store #5; by clue 14, then, the artwork shop would be Gail's, while the bath shop would be Fran's. By clue 4, stores #2, #4, #6, and #8 would be, in some order, Bill's, the linen shop, the basket shop, and the kitchenware shop. This, however, would leave no place for Karen's Light Fixtures (clue 11); Ian's is not store #1. If Ian's was store #3, the bath shop would be store #1, Art's Appliances would be store #10, and Heidi's would be store #12. By clue 8, then, Ellen's would be the candle shop in #7, and Don's would be store #11. Again, this would leave no location for Jack's (clue 4), so Ian's is not store #3. If Ian's was store #12, then Art's Appliances would be store #1 and the bath shop would be store #3. However, two shops must separate Art's Appliances and the bath shop, so Ian's is not store #12. Ian's, then, is store #10, the bath shop is store #12, Art's Appliances is store #3, and Heidi's is store #1. Jack's is one of the outer shops, and it does not connect to Ian's, Heidi's, or Art's (clue 4); it is store #9. The glassware shop, then, is either store #6 or store #8 (clue 4). If it were store #8, then Fran's and Gail's would be, in some order, #12 and #6 (clue 4). By clue 2, then, the artwork shop would be store #5, and the glassware shop would be Bill's. This, however, would contradict clue 13; the glassware shop is not #8, so it is #6. That shop, then, is Don's, and the candle shop is store #2 (clue 7). Since the artwork shop does not connect to Don's Glassware, it is store #8 (clue 2). Fran's and Gail's are, in some order, the bath shop and the artwork shop (clue 4). Fran's is not the artwork shop (clue 14), so Gail's is and Fran's sells bath goods. Ellen's sells linens, baskets, or kitchenware (clue 2). It does not sell baskets or kitchenware (clue 8); it sells linens. Karen's sells light fixtures (clue 11). It does not connect to Gail's Artworks (clue 2), so it is store #4. By clue 3, Leigh's is not store #11 or #5; it is store #2, which sells candles. Heidi's, then, is either the Christmas or the rug shop (clue 3); it is not the rug shop (clue 12), so it is the Christmas shop. Store #5, then, is the rug shop (clue 3). By clue 2, then, the rug shop is Bill's. By elimination, store #11 is Carole's. Carole's and Jack's sell, in some order, basketry and kitchenware (clue 2); by elimination, Ian's sells pottery. Carole's does not sell basketry (clue 5); it sells kitchenware and, by elimination, Jack's sells basketry. In summary:

1: Heidi's Christmas Shop
2: Leigh's Candles
3: Art's Appliances
4: Karen's Lighting
5: Bill's Rugs
6: Don's Glassware
7: Ellen's Linens
8: Gail's Artwork
9: Jack's Baskets
10: Ian's Pottery
11: Carole's Kitchenware
12: Fran's Bath Shop

73. JURY DUTY

The first three people on the list were accepted (clue 1), so they sat in jury seats #1, #2, and #3. Catherine was #8 on the list, and Mr. King was #16 (clue 10). The person who was #21 on the list was accepted for jury seat #12 (clue 1). There were five men on the jury, two in the front row and three in the back (clue 15). By clue 20, Henry and Mr. Rothman sat next to each other in the front row. By clue 2, Clarence sat in the front row, so he is Mr. Rothman. Since Mr. Glover sat in the front row (clue 18), he is Henry. The electrician sat directly behind Clarence Rothman (clue 2). Ms. Waters sat between Clarence Rothman and Catherine on the jury (clue 16). Margaret sat between the realtor and Henry Glover (clue 18). By clue 20, then, there are two possible seating arrangements: that the front row was, in order from left to right, the realtor, Margaret, Henry Glover, Clarence Rothman, Ms. Waters, and Catherine, while the electrician sat behind Clarence Rothman and Jerome sat

169

behind Ms. Waters, or the front row was, in order from left to right, Catherine, Ms. Waters, Clarence Rothman, Henry Glover, Margaret, and the realtor, while Jerome sat behind Ms. Waters and the electrician sat behind Clarence Rothman. Catherine, then, would be in jury seat #1; she would have been #1 on the list (clue 1). Since she was #8 on the list, the correct order is: front row, the realtor in #1, Margaret in #2, Henry Glover in #3, Clarence Rothman in #4, Ms. Waters in #5, and Catherine in #6, while in the back row the electrician sat in #10 and Jerome in #11. By clue 15, Richard (clue 3) sat in the back row, therefore Catherine is the hotel clerk, while Richard sat in #7. The first three people on the list were in jury seats 1, 2, and 3 (clue 1); they were, in order, the realtor, Margaret, and Henry Glover. The first man on the list is the auto mechanic, and the first woman on the list was Alice (clue 6); the realtor, then, is Alice, while Henry Glover is an auto mechanic. By clue 3, then, Iris was in seat #12, and realtor Alice is named Johnson. Iris was #21 on the list (clue 1), so her last name is Channing (clue 6). Since she was last on the list and accepted for the jury, she was the last accepted on the jury: she is a nurse (clue 8). On the jury, Richard sat directly behind Alice Johnson, the electrician sat directly behind Clarence Rothman, Jerome sat directly behind Ms. Waters, and Iris Channing sat directly behind Catherine. By clue 2, then, Ruth sat in seat #8, directly behind Margaret Olson in seat #2, while Mr. Friedman is Jerome in seat #11, and Gloria is Ms. Waters in seat #5. By clue 12, then, Ms. Young sat in #9, Robert in #10 and is the electrician, and the waiter in #11 and is Jerome. Three of the jurors are, in order, the antiques dealer, Lovell, and the man who is a musician (clue 11); the only possibility is that Catherine is Lovell, Richard is the musician, and Gloria Waters is the antiques dealer. At this point, on the list we have 1: Alice Johnson, realtor, accepted, seat #1; 2: Margaret Olson, accepted, seat #2; 3: Henry Glover, auto mechanic, accepted, seat #3; 8: Catherine Lovell, hotel clerk, accepted, seat #6; 16: Mr. King; 21: Iris Channing, nurse, accepted, seat #12. There are a total of ten women on the list, and four are already placed. By clue 14, four of the women were listed consecutively, followed by Larry, followed by three other women; one of the women already placed on the list must be in either the group of four women before Larry, or in the group of three women after Larry. Since Mr. King is #16, Iris Channing is not in either group. Since Henry Glover is #3, neither Alice Johnson nor Margaret Olson are in either group. Catherine Lovell, then, is in the group of four women before Larry. Three of the people on the list consecutively were the chiropractor, Norman, and Gloria Waters (clue 9). By clue 16, then, Norman is Mr. Simmons, and the order consecutively was: the chiropractor, Norman Simmons, Gloria Waters, and Ms. North. Gloria Waters and Ms. North, then, must be the first two of the four women preceding Larry. Since Henry is the auto mechanic, the chiropractor was #4 on the list, Norman Simmons #5, Gloria Waters #6, and Ms. North #7. Catherine Lovell was, then, the third woman of the four preceding Larry: a woman was #9, and Larry #10, while the remaining women were #11, #12, and #13. The remaining men, then, were #4, #14, #15, #16, #17, #18, #19, and #20. By clue 6, Anthony was #20. Norman Simmons was not on the jury, so he was dismissed either immediately or after others had been questioned. If Ms. North, who was on the list between Gloria Waters and Catherine Lovell, had held a seat temporarily, there would have been a final jury member between Gloria and Catherine; since they sat next to each other on the jury, Ms. North was dismissed immediately. By clue 7, then, Mr. Braden appeared on the list before Ms. North; he is the chiropractor, and was dismissed immediately. Jennifer was second to be dismissed (clue 7), so she is Ms. North, and Norman Simmons was temporarily accepted in seat #4, then dismissed and replaced by a person appearing later on the list. Catherine Lovell was #8 on the list, and sat in seat #6 on the jury, while Richard sat in seat #7. The person who was #9 on the list was a woman, while Larry was #10, and #11, #12, and #13 are the remaining women. Either the woman who was #9 on the list or Larry, then, was temporarily accepted for seat #7, then dismissed later and replaced by Richard, while the other was dismissed immediately. By clue 17, Ms. Fox was #11 on the list, the housewife #12, and Susan #13. By clue 19, the dietitian was #7 or #9, the computer programmer #9, #10, or #11, and Ms. Potter was #13, and her first name is Susan. Since Ms. Young was accepted for the jury, she was not the woman listed #9; she was the woman listed #12, the housewife. Since the woman listed #9 was dismissed, Ruth was listed #11. By elimination, Ms. Thompson was the woman listed #9. Ms. Thompson and Ms. Young are, in some order, Patricia and Dorothy. Dorothy was dismissed (clue 4), so she is not Ms. Young; Patricia is Ms. Young and Dorothy is Ms. Thompson. By clue 5, then, Sue Potter is the teacher, while Moore was listed #14. Teacher Sue Potter was not on the final jury, so she was dismissed; Patricia and Moore, then, were both accepted (clue 5). Since Jerome Friedman was in seat #11 and Iris Channing in seat #12, Moore was beside Ms. Young on the jury; he is Robert, the electrician in seat #10. Sue Potter, then, was dismissed immediately. Dorothy Thompson was accepted temporarily for seat #7 (clue 4),

170

so Larry was dismissed immediately. He was, then, the third dismissed, so he is the jeweler (clue 7). The only final jury members whose occupations are not yet known are Margaret Olson, #2, Clarence Rothman, #4, and Ruth Fox, #8; by clue 13, Ruth is the accountant. Margaret is not the photographer (clue 8); Clarence is. Margaret, then, is the florist (clue 4). By clue 19, the dietitian is Jennifer North, and Dorothy Thompson is the computer programmer. Mr. Braden was #4 on the list, and he was dismissed immediately. He is not Jerome, Clarence, or Richard, who were on the final jury, nor is he William (clue 21), or Frank (clue 22); he is Michael. Mr. Dreyer was accepted for the jury (clue 8); he is Richard. The last names of the remaining men who were dismissed are Varner, King, who was #16 on the list, Wilkins, and Quigley. The remaining first names are Larry, who was #10, William, Frank, and Anthony, who was #20. Neither Larry nor Anthony, then, is Mr. King; either Frank or William is. Frank is not Mr. King (clue 22), so William is. Wilkins, then, was #17 (clue 21). Frank was dismissed after two others had been questioned, and Quigley was dismissed after that (clue 22); Frank is Mr. Wilkins, and Anthony is Quigley. By elimination, Larry is Varner. Robert Moore was #14 on the list and took seat #10 in the final jury. The man who was #15 on the list was accepted, so he was Jerome Friedman, who had seat #11 on the final jury. By elimination, Frank Wilkins was temporarily accepted for seat #12. William King, then, who preceded Frank on the list, was dismissed immediately. All twelve jury seats have now been filled; Clarence Rothman, Richard Dreyer, and Iris Channing, in order, replaced those who were accepted temporarily. Clarence, who took seat #4, replaced Norman Simmons; Richard Dreyer, who took seat #7, replaced Dorothy Thompson, and Iris Channing, who took seat #12, replaced Frank Wilkins. Since they replaced in consecutive order, Clarence was #18 on the list, and Richard was #19. The office manager was listed after Wilkins (clue 21); he is Anthony Quigley, #20. The veterinarian was dismissed immediately (clue 4), so he is not Norman Simmons or Frank Wilkins; he is William King. The bus driver was dismissed after at least five others were questioned (clue 4); he is not Frank Wilkins, who was dismissed after two others were questioned (clue 22); he is Norman Simmons. By elimination, Frank is the carpenter. In summary:

1. Alice Johnson (#1), realtor
2. Margaret Olson (#2), florist
3. Henry Glover (#3), auto mechanic
4. Michael Braden, chiropractor
5. Norman Simmons, bus driver
6. Gloria Waters (#5), antiques dealer
7. Jennifer North, dietitian
8. Catherine Lovell (#6), hotel clerk
9. Dorothy Thompson, computer programmer
10. Larry Varner, jeweler
11. Ruth Fox (#8), accountant
12. Patricia Young (#9), housewife
13. Susan Potter, teacher
14. Robert Moore (#10), electrician
15. Jerome Friedman (#11), waiter
16. William King, veterinarian
17. Frank Wilkins, carpenter
18. Clarence Rothman (#4), photographer
19. Richard Dreyer (#7), musician
20. Anthony Quigley, office manager
21. Iris Channing (#12), nurse

74. BASKETBALL FIVE

Each player scored a different number of two-point goals, and each scored at least 1 (clue 8). Bob is not Rowe (clue 5), so the pairs mentioned in clue 1 are four different players. By clue 1, then, if Bob had scored 1 two-point goal, Smith would have scored 2, Rowe would have scored at least 3, and Chuck at least 6. The total number of two-point goals, then, for all five players would have been at least 16 (1+2+3+4+6), i.e., 32 points. There were, then, at least 12 one-point goals (clue 7). Since the total number of points scored in the game was 71, the most that could have been scored in three-point goals was 27. Bob scored the most three-point goals: two more than the player who scored the least (clue 5). To reach a total of no more than 9 three-point goals, the only possibility is that Bob scored 2 three-point goals, the player who scored the least scored none, and the other three players scored one each, for a total of 5 three-point goals. The total number of points earned in one- and two-point goals, then, was 56. By clue 7, the only possibility is that there were a total of 20 two-point goals and 16 one-point goals. For five players, the average of 71 points is 14.2. Exactly two players, then, scored 14 or 15 points (clue 2); the other players scored more than 15 or less than 14. By clue 10, then, if both Al and Phelps had scored 14, a third player would have scored either 14 or 15 to make the total 71. If Al and Phelps scored, in some order, 14 and 15 points, the five scores were either 13, 13, 14, 15, 16 or 12, 13, 14, 15, 17 (clue 10). If both Al and Phelps scored 15 points, the scores were either 11, 13, 15, 15, 17 or 12, 13, 15, 15, 16 (clue 10). The lowest score, then, was either 11, 12, or 13; the second lowest score was 13, the third lowest score was either 14 or 15, the second-highest score was 15, and the highest score was either 16 or 17. One guard scored 11, 12, or 13, the second guard scored 13, and a forward scored either 16 or 17 (clue 4). The center and the second forward, then, scored 14 or 15 points; they are, in some order, Al and Phelps (clue 2). Phelps is not the center (clue 2), so he is the second forward, and Al is the center. One forward scored the same number of one- and two-point goals (clue 6); his total score, then, must be divisible by three. Since the range of total scores is from 11 to 17, he scored either 12 or 15 points; since only a guard could have scored a 12-point total (clue 4), the forward scored 15. That forward, then, is Phelps. Each player scored a different number of two-point goals, and each scored at least one (clue 8). Since there was a total of 20 two-point goals scored, by clue 1, there are three possible sets of two-point scores: 1, 2, 3, 6, 8; 1, 2, 4, 5, 8; 2, 3, 4, 5, 6. Bob scored 1, 2, 3, or 4 two-point goals (clue 1). Bob did not score 2 two-point goals (clue 6). If a player scored 8 two-point goals he would have scored no three-point goals and 1 one-point goal (clue 9), for a total of 17 points, the maximum. If Bob had scored 4 two-point goals, he would have scored a total of at least 15 points, Smith would have scored 17 points, Chuck would have scored 2 two-point goals and 1 three-point goal, and Rowe would have scored 1 two-point goal, and 1 three-point goal (clue 1). This, however, contradicts clue 6. Bob, then, did not score 4 two-point goals. If the five two-point totals were 1, 2, 4, 5, 8, by clue 1, Bob would have scored 1 two-point goal, Chuck would have scored 8 two-point goals, 1 one-point goal, and no three-point goals for a total of 17, Smith would have scored 2 two-point goals and 1 three-point goal, and Rowe would have scored 4 two-point goals and 1 three-point goal. Since Chuck would have scored 1 one-point goal, Wolfe would have scored 1 two-point goal (clue 3); Bob would be Wolfe. Phelps would not be Chuck (clue 2). By elimination, then, Phelps would be the fifth player, who scored 5 two-point goals and 1 three-point goal. That accounts for 13 points. He did not score only 1 one-point goal (clue 6), and there is no combination allowing one 16 point score and one 17 point score; he would have made 2 one-point goals for a total score of 15. Rowe, who would have scored 4 two-point goals, could not be the player who scored the same number of one- and two-point goals, or his total would also be 15. Bob Wolfe could not be that player, or he would have scored only 9 points. Smith, could not be that player, or he would have scored only 9 points. The scores for two-point goals, then, were not 1, 2, 4, 5, 8. If the scores were 1, 2, 3, 6, 8, either Bob or Rowe scored 1 two-point goal. If Bob scored 1, then Smith scored 2, Rowe scored 3, Chuck scored 6, and the fifth player scored 8. That fifth player would have scored 1 one-point goal, 8 two-point goals, and no three-point goals for a total of 17. Chuck would have scored at least 1 one-point goal, 6 two-point goals, and 1 three-point goal for a total of at least 16. However, there is no combination of scores allowing one 16 point score and one 17 point score; Bob did not score 1 two-point goal. If Rowe had scored 1 two-point goal, by the same reasoning, Smith would have scored at least 16 points and the fifth player would have scored 17; the two-point scores were not 1, 2, 3, 6, 8. The two-point scores, then, were 2, 3, 4, 5, 6; Rowe scored 2 two-point goals, Bob scored 3, Chuck scored 4, Smith scored 6, and the fifth player scored 5. Chuck is not Wolfe (clue 3). If Bob were Wolfe, Chuck would have scored 3 one-point goals (clue 3); his total score would have been 11 or 14 points. Only Al could have gotten 14 points, however (clue 2), so Chuck's total would have been 11, and he would be the player who scored no three-point goals. Since Phelps scored 15 points, Chuck would be Tate. Phelps, by elimination, would be the player who scored 5 two-point goals; since he would have scored 1 three-point goal as well, and since his total score was 15, he would have scored 2 one-point goals. Since Chuck Tate would have scored 11 points, the only possibility would be that the five scores would have been 11, 13, 15, 15, 17; Phelps and Al would have both scored 15 points. Smith scored 6 two-point goals. If Chuck Tate had scored no three-point goals, Smith would have scored 1 (clue 5), and at least 1 one-point goal (clue 9); his total score would be at least 16, so he would have been the player who scored 17. He would have scored 2 one-point goals. Al, who scored 15 points, would then be Rowe. Since

172

Rowe scored 2 two-point goals and would have scored 1 three-point goal, he would have scored 8 one-point goals. By elimination, then, Bob Wolfe would have scored 13 points; he would have scored 1 one-point goal, 3 two-point goals, and 2 three-point goals. This, however, would leave no player with the same number of one- and two-point goals, contradicting clue 6. Bob, then, is not Wolfe; Wolfe is the fifth player, who scored 5 two-point goals. By clue 3, then, Chuck scored 5 one-point goals. His total score, then, was either 13 or 16 points; he is not Phelps, who scored 15 points, so he is Tate. Bob, by elimination, is Phelps. Since forward Bob Phelps scored a total of 15 points, he scored 3 one-point goals. Ed scored 4 one-point goals (clue 3). Rowe scored 2 two-point goals. Since the minimum score was 11, he scored 1 three-point goal, and at least 4 one-point goals. If he were not Ed, the minimum total number of one-point goals would have been Rowe's 4, Bob Phelps' 3, Chuck Tate's 5, Ed's 4, and the fifth player's 1: a minimum total of 17. Since there were only 16 one-point goals scored in all, Ed is Rowe, and he scored 4 one-point goals, 2 two-point goals, and 1 three-point goal for a total score of 11. From the possible score combinations, the only one that contains a score of 11 is 11, 13, 15, 15, 17; those were the first players' scores. Ed Rowe is the guard who scored the lowest (clue 4). Chuck Tate did not score 16 points, so he scored 13 points, and scored no three-point goals. He is the guard who scored the second-lowest (clue 4). Smith scored 6 two-point goals, 1 three-point goal, and at least 1 one-point goal (clue 9); he scored at least 16 points. He, then, is the forward who scored 17 points (clue 4); he scored 2 one-point goals, 6 two-point goals, and 1 three-point goal. He is not, then, Al, who scored 15 points; he is Dan. By elimination, Wolfe is Al. He is the center who scored 15 points; since he scored 5 two-point goals and 1 three-point goal, he scored 2 one-point goals. In summary, with goals listed in order from one-point to three-point:

Guard Ed Rowe; 4, 2, 1: 11
Guard Chuck Tate; 5, 4, 0: 13
Forward Bob Phelps; 3, 3, 2: 15
Center Al Wolfe; 2, 5, 1: 15
Forward Dan Smith; 2, 6, 1: 17

75. FARMERS' MARKET

Noreen is not Ms. Thompson (clue 5), Quincy, Purcell, or Rush (clue 6); she is Ms. Smith, Janet is not Ms. Quincy, Purcell, or Rush (clue 6); she is Ms. Thompson. Neither Karen nor Loretta is Ms. Quincy (clue 3); Marcia is. Karen is not Ms. Purcell (clue 10); Loretta is and, by elimination, Karen is Ms. Rush. Three vendors sold apples (clue 3), two sold walnuts (clue 4), two sold squash (clue 11), two sold carrots (clue 13), two sold almonds (clue 14), two sold tomatoes (clue 17), and two sold cantaloupes (clue 19). Since each vendor sold two types of produce (clue 1), one vendor sold each of the following: corn, raspberries (clue 5), yellow onions (clue 7), lettuce (clue 8), watermelons, pumpkins (clue 9), greens beans (clue 13), green onions (clue 15), and strawberries (clue 16). Vendor #3 sold apples (clue 3) and walnuts (clue 4). Vendors #6 and #8 also sold apples (clue 3). Vendor #7 sold walnuts (clue 4), Vendor #1 sold corn (clue 5). The vendor who sold pumpkins was at #10 or #12 (clue 15). That vendor had at least two customers (clue 9). The vendor at #10 had only one customer (clue 12), so the pumpkin vendor was at #12. Vendor #1, then, sold green onions (clue 15) and corn. The vendor at #7, who sold walnuts, did not sell lettuce (clue 8), watermelons (clue 9), squash (clue 11), green beans, yellow onions, or carrots (clue 13), tomatoes (clue 17), cantaloupes (clue 19), strawberries, or raspberries (clue 22); that vendor sold walnuts and almonds. There were twelve pairs of produce sold (clue 1). One pair was corn and green onions, a second pair was apples and walnuts, a third was walnuts and almonds, the fourth and fifth were green beans and carrots and yellow onions and carrots (clue 13). A sixth pair was sold by the second almond vendor, who was at a corner table (clue 14). The two squash vendors were not at corner tables (clue 11); they sold the seventh and eighth pairs. There were two cantaloupe vendors, neither of whom was at a corner table (clue 19). Neither sold squash in addition to cantaloupes (clue 11); they sold the ninth and tenth pairs. There were two tomato vendors, who were not at corner tables (clue 17) so they didn't sell almonds (clue 14). They did not sell squash (clue 11) or cantaloupes (clue 19); they sold the eleventh and twelfth pairs. The second and third apple vendors were at sites #6 and #8 (clue 3). The squash vendors were at #2 and #12, #6 and #8, or #9 and #11 (clue 11). If they were at #6 and #8, both would have sold apples and squash, contradicting clue 1; neither squash vendor sold apples as their second item. The second almond vendor was at a corner table (clue 14), and so did not sell apples. By clue 1, then, one apple vendor also sold cantaloupes, while the other also sold tomatoes. The apples and tomatoes were not sold at #6 (clue 17); they were sold at #8, while apples and cantaloupes were sold at #6. Tomatoes and cantaloupes and tomatoes and squash were sold at adjoining tables (clue 17), and none were sold at corner tables. Tomatoes were not sold at #5 (clue 17). If squash were sold at #9, it was also sold at #11 (clue 11). That leaves no place for the second tomato vendor to be next to the second cantaloupe vendor (clues 11, 17). The second cantaloupe vendor was at #9. By clue 11, then, squash was sold at #2 and #12, where the combination was squash and pumpkins. By clue 17, then, the second tomato vendor was at #11. The cantaloupe vendor at #9 did not sell yellow onions (clue 7), strawberries (clue 16), raspberries, or lettuce (clue 19); he sold watermelons. The produce sold with lettuce was not squash or almonds (clue 8); it was tomatoes, and they were sold at #11. To recap what we have discovered so far: corn and green onions were sold at #1, squash was sold at #2, apples and walnuts were sold at #3, apples and cantaloupes were sold at #6, walnuts and almonds were sold at #7, apples and tomatoes were sold at #8, cantaloupes and watermelons were sold at #9, tomatoes and lettuce were sold at #11, and pumpkins and squash were sold at #12. By clue 13, yellow onions and carrots were sold at #4 or #5, and green beans and carrots were sold at #5 or #10. By clue 14, almonds were sold at #4 or #10. By elimination, strawberries and raspberries were sold, in some order, with squash at #2 and almonds at #4 or #10. By clue 2, each buyer bought at least five items and at most seven. By clue 6, then Karen Rush bought seven items, Janet Thompson and Noreen Smith each bought 6, and Loretta Purcell and Marcia Quincy each bought five. Since each purchaser shopped at exactly five tables (clue 2), Karen Rush bought the two items offered at two tables and one item each at three, Noreen and Janet each bought the two items offered at one table and one each at four others, while Marcia and Loretta each bought one item from five different tables. Karen, Loretta, and Marcia bought apples (clue 3). The vendor at #6 had only one customer (clue 12). Since each vendor sold both items (introduction), the one customer at #6 bought both apples and cantaloupes. Since Loretta and Marcia bought five items from five different vendors, they did not buy apples at #6; Karen Rush bought apples there, as well as cantaloupes. Loretta Purcell did not buy her apples at #3 (clue 10); she bought apples at #8 and Marcia bought them at #3. Marcia and Noreen bought walnuts (clue 4). Since Marcia Quincy did not buy two items from the same vendor, she did not buy walnuts from #3; she bought them from #7, and Noreen bought them from #3. Noreen and Janet bought corn at #1 (clue 5). The woman who bought the watermelon at #9 also bought both nuts and berries (clue 9). Karen Rush and Janet Thompson bought neither walnuts (clue 4) nor almonds (clue 18), they did not buy the watermelon. Noreen Smith bought neither raspberries (clue 5) nor strawberries (clue 18); she did not buy the watermelon. Marcia Quincy did not buy the watermelon (clue 9). By elimination, Loretta Purcell bought the watermelon at #9. She did not buy walnuts (clue 4), so she bought almonds (clue 9). Noreen did not buy strawberries (clue 18), so she also shopped at #9 (clue 16); she bought cantaloupes there, while Karen, Marcia, and Janet all bought strawberries (clue 16). Karen Rush shopped at three corner tables (clue 6); since she bought neither walnuts (clue 4) nor almonds (clue 18), she did not shop at #7, so she shopped at #1, #4, and #10. Since #10 had only one customer (clue 12), she bought both items from that table. Almonds, then, were not sold at #10; they were sold at #4 (clue 14). By clue 13, then, green beans and carrots were sold at #10, and Karen bought both, while yellow onions and carrots were sold at #5. Almonds were sold at #4 and #7. Loretta Purcell, who bought almonds and a watermelon, did not shop at #7 (clue 9); she bought her almonds at #4. Loretta bought the watermelon, so she bought at least one type of berry (clue 9); she did not buy strawberries (clue 16), so she bought raspberries. Since she bought almonds at #4, she did not buy raspberries there (clue 2). Strawberries were sold at #4 with the almonds, while raspberries were sold to Loretta at #2 with the squash. Karen, Marcia, and Janet all bought their strawberries at #4. #5 had only one customer (clue 12), so that customer bought both items offered. That customer was not Loretta or Marcia. Since Karen bought both items at #6 and both items at #10, she did not shop at #5. Noreen bought cantaloupe at #9, so she did not shop at #5 (clue 7); Janet Thompson bought both yellow onions and carrots at #5. Janet and Karen did not buy almonds (clue 18). Since Marcia bought walnuts at #7, she did not buy almonds there; Noreen Smith bought the almonds at #7. One woman bought a pumpkin, a nut, and a berry (clue 9). That woman was not Karen or Janet, who bought neither walnuts (clue 4) nor almonds (clue 18), Noreen, who bought neither raspberries (clue 5) nor strawberries (clue 18), or Loretta (clue 9); Marcia bought a pumpkin at #12. There were two customers at #11 (clue 20). They were not Karen (clue 20), Loretta (clue 9), or Janet (clue 7); they were Marcia and Noreen. Since Marcia bought the pumpkin at #12, she did not buy lettuce at #11 (clue 8); as her final purchase, she bought tomatoes at #11, and Noreen bought lettuce. Noreen bought corn at #1, walnuts at #3, cantaloupe at #9, almonds at #7, and lettuce at #11. She bought two items from one vendor (clue 2). She did not buy apples at #3 (clue 3), watermelon at #9 (clue 9), or walnuts at #7 (clue 2), so her final purchase was either green onions at #1 or tomatoes at #11. Two women bought squash (clue 11); Karen did not (clue 20), so Loretta and Janet both did. Loretta bought squash, her final purchase, at #12, and Janet bought squash at #2. Janet, then, bought tomatoes (clue 21) as her final purchase. She did not buy them from #11 (clue 7); she bought them from #8. By elimination, Karen Rush bought the second pumpkin (clue 15) from #12. Since she made a purchase from #1, she was not the third tomato buyer (clue 21); Noreen bought tomatoes from #11. By elimination, Karen's final purchase was green onions from #1. In summary:

- #1: corn (Noreen Smith, Janet Thompson) and green onions (Karen Rush)
- #2: squash (Janet) and raspberries (Loretta Purcell)
- #3: apples (Marcia Quincy) and walnuts (Noreen)
- #4: almonds (Loretta) and strawberries (Karen, Marcia, Janet)
- #5: yellow onions and carrots (Janet)
- #6: apples and cantaloupes (Karen)
- #7: walnuts (Marcia) and almonds (Noreen)
- #8: apples (Loretta) and tomatoes (Janet)
- #9: cantaloupes (Noreen) and watermelons (Loretta)
- #10: green beans and carrots (Karen)
- #11: tomatoes (Marcia, Noreen) and lettuce (Noreen)
- #12: pumpkins (Karen, Marcia) and squash (Loretta)